GEOGRAPHERS
Biobibliographical
Studies

VOLUME 20

GEOGRAPHERS: BIOBIBLIOGRAPHICAL STUDIES

This volume is part of a series of works on the history of geography planned by the Commission on the History of Geographical Thought of the International Geographical Union and the Commission of the International Union of the History and Philosophy of Science. *Chair:* Professor Vincent Berdoulay, Département de Géographie, Université de Pau, rue de Doyen Poplawski, 64000 Pau, France. *Secretary:* Dr Mark Bassin, Department of Geography, University College London, UK. *Other Full Members:* Dr Patrick Armstrong, Department of Geography, University of Western Australia, Nedlands, Western Australia 6907 (also *Co-Editor: Geographers: Biobibliographical Studies*); Dr Ahmed Bencheikh, Département de Géographie, Université Cadi Ayyad, B.P.S., 17, Quartier Amerchich, Marrakech, Morocco; Dr Athanase Bopda, Cameroon; Professor Gary Dunbar, 13 Church Street, Cooperstown, New York 13326, USA; Professor Josefina Gomez Mendoza, Spain; Dr Lia Osorio Machado, Brazil; Dr Hideki Nozawa, Japan; Dr Ute Wardenga, Institut für Länderkunde, Schongauerstrasse 9, 04329 Leipzig, Germany; Dr Hong-Key Yoon, Department of Geography, University of Auckland, Auckland, New Zealand. *Honorary Members:* Professor Anne Buttimer, Department of Geography, University College Dublin, Belfield, Dublin 4, Ireland; Professor David Hooson, Department of Geography, University of California, Berkeley, California 90047, USA; Professor Philippe Pinchemel, Centre de Géohistoire, 7 rue Malher, 75007 Paris, France; Professor Keiichi Takeuchi, Department of Geography, Komazawa University, Setagayaku, Tokyo 154, Japan; Professor Oskar H.K. Spate, Research School of Pacific Studies, Australian National University, GPO Box 4, Canberra, ACT 2601, Australia; Professor Johannes A. van Ginkel, Fakultiet der Ruimtelijke Wetenskappen, Rijksuniversiteit te Utrecht, Heidelberglaan 8, P.B. 80125, 3508 TC Utrecht, The Netherlands. *Co-Editor: Geographers: Biobibliographical Studies:* Professor Geoffrey J. Martin, 33 Fairgrounds Road, Woodbridge, Connecticut 06525, USA.

GEOGRAPHERS
Biobibliographical Studies

VOLUME 20

Edited by Patrick H. Armstrong
and Geoffrey J. Martin

on behalf of the
Commission on the History of Geographical Thought
of the International Geographical Union and the
International Union of the History and Philosophy of Science

CONTINUUM
London and New York

First published 2000 by
Continuum
Wellington House, 125 Strand, London WC2R 0BB
370 Lexington Avenue, New York, NY 10017-6503

© International Geographical Union 2000

All rights reserved. No part of this publication may be reproduced or transmitted in any form or by any means, electronic or mechanical, including photocopying, recording or any information storage or retrieval system, without permission in writing from the publishers or their appointed agents.

British Library Cataloguing-in-Publication Data
Geographers: biobibliographical studies — Vol. 20
 1. Geographers — Biography — Periodicals
 910'92'2   G67

ISBN 0-8264-4960-3

Typeset by BookEns Ltd, Royston, Herts.
Printed and bound in Great Britain by Biddles Ltd, Guildford and King's Lynn

Contents

The Contributors		vi
List of Abbreviations		vi
Introduction	Patrick H. Armstrong and Geoffrey J. Martin	vii
John Adair 1660–1718	Charles W. J. Withers	1
James Cook, RN 1728–1779	Jill Rutherford and Patrick H. Armstrong	9
Owen Lattimore 1900–1989	Gary S. Dunbar	24
Tsunesaburo Makiguchi 1871–1944	Keiichi Takeuchi	43
Thomas Robert Malthus 1766–1834	Patrick H. Armstrong	57
Akira Nakanome 1874–1959	Hiroshi Ishida	68
John Ogilby 1600–1676	Charles W. J. Withers	77
Thomas Pennant 1726–1798	Colin Thomas	85
Charles-Eugène Perron 1837–1909	Peter Jud	102
Pedro C. Sánchez Granados 1871–1956	Héctor Mendoza Vargas	108
Zheng He 1371–1433	Lu Zi and Liu Yang	119
Index		126

The Contributors

Patrick Armstrong teaches geography at the University of Western Australia and is a co-editor of *Geographers*.

Gary Dunbar is Professor Emeritus of Geography at the University of California, Los Angeles, and now lives in Cooperstown, New York.

Hiroshi Ishida is Professor Emeritus of Geography at the University of Hiroshima.

Peter Jud teaches geography in a high school near Zurich.

Liu Yan is a professor in the Department of Geography, Hebei Teachers' University, People's Republic of China.

Lu Zi is a professor in the Department of Geography, Hebei Teachers' University, People's Republic of China.

Hector Mendoza Vargas is at the Institute of Geography, National University of Mexico.

Jill Rutherford works for the International Baccalaureate Organisation and the University of Hull.

Keiichi Takeuchi is Professor of Geography at Komazawa University, Tokyo.

Colin Thomas is Reader in Geography at the University of Ulster, Coleraine.

Charles W. J. Withers is Professor of Geography at the University of Edinburgh.

List of Abbreviations

IPR	Institute of Pacific Relations
N.G.U.	Elisée Reclus, *Nouvelle Géographie Universelle*
OAS (in Spanish, OEA)	Organization of American States
PAIGH	Pan-American Institute of Geography and History
RGS	Royal Geographical Society
SFH	Owen Lattimore, *Studies in Frontier History*

Introduction

The publication of the twentieth volume of *Geographers: Biobibliographical Studies* provides an appropriate opportunity for the re-examination of the aspirations that were held for the series when the first volume appeared in 1977. In the Introduction to the first volume, Professor Philippe Pinchemel, then the Chairman of the Commission on the History of Geographical Thought of the International Geographical Union, wrote as follows:

> Until recently ... the history of geography has been limited to the history of the discovery and exploration of the earth ... But in fact the history of the recording of the earth's surface has become the history of one particular branch of human learning – geography – as understanding of the earth has deepened. The history of geography may be conceived as the history of the earth's surface as it progressively became the habitat of human expansion, through the gradual humanization of the earth. It is the history of the changing relationships between environments, space and territories on the one hand, and societies, cultures and civilizations on the other, as told through the description, analysis and explanation of human enterprise within its setting of natural environments. It reveals how geography as a science has been transformed from the study of locations to the study of earth sculpture, and thence to one of natural environments and lastly of social structures within those environments.
>
> These successive geographical sciences have been conceived by minds with a diversity of training – by cartographers, philosophers, natural historians and political scientists – who had one thing in common, a continuing curiosity about the variety of the earth's surface forms and of the character of its human development.

Later in the same essay, while suggesting that geography in the nineteenth century was dominated by four names – Humboldt, Ritter, Ratzel and Vidal de la Blache – he noted that alongside these great names:

> ... there are many others less well known but without whom the whole development of geography cannot be appreciated. Among them were university men, explorers, statisticians, men who founded and worked for geographical societies, map curators ... historians, natural historians and geologists. For geography has never been the exclusive concern of geographers ...

The brief that was set forth in 1977 has been closely adhered to. There is perhaps a slightly Eurocentric flavour to some of the above, but Professor Pinchemel emphasized the international nature of the enterprise upon which the Commission was embarking. In the succeeding years there have been included geographical figures from many countries and all continents, and the authorship has been similarly international. Also, despite the phrase 'university men' above, and the phrase 'man and ... his environment' elsewhere in the introductory essay, a number of accomplished women geographers have been described.

But openness has always been a characteristic of the volumes. Successive editors have always attempted to achieve a balance between the titans and those lesser figures who have given character to the subject. The test has been whether a person to be included in the pages of *Geographers* made a significant contribution to the development of geographical thought.

The present volume in many respects includes a typical selection of the type of personages envisaged at the commencement of the series. The lives of the individuals reviewed extend from the fourteenth century to the twentieth; three come from Asia, six from Europe, and two from North America. They include figures of international importance and those whose contributions have been more of local significance.

Only a minority of those represented would have referred to themselves as geographers in the strictest sense, and indeed several of the categories envisaged in the introductory essay are included – the explorer, the historian (perhaps antiquarian might be more accurate for one or two of the subjects), the statistician or political scientist, the naturalist, the cartographer.

Some of those who have made a contribution to geography have been polymaths, and the subject of the first essay in this volume is such a person. John Adair (1660–1718) has been variously described as a mathematician, map-maker, natural historian and antiquarian. He made both terrestrial and ship-based coastal surveys of Scotland, and held the titles of Queen's Geographer and Geographer for Scotland. His cartographic work is important in revealing 'how the early modern state turned to maps to "make space" for itself as a political and economic entity'.

The towering figure of Captain James Cook, RN (1728–1779) follows. He was also a man of wide interests, which included ethnography, linguistics and astronomy. He was an explorer in the traditional sense, a hydrographic surveyor and an enlightened leader of men. Of him it has been said 'the map of the Pacific was his ample panegyric'. It was he who disposed, once and for all, of the myth of the 'Great South Land'.

Owen Lattimore (1900–1989) was an American, although he lived a large part of his life outside the United States. He worked as a journalist as well as a university teacher (at Johns Hopkins University and the University of Leeds), and was an authority on the geography and cultural history of Inner Asia and China. He developed an excellent knowledge of the languages necessary for the prosecution of his research, but he is perhaps best remembered for his persecution by extreme right-wing politicians in the early years of the Cold War.

Tsunesaburo Makiguchi (1871–1944) had a significant effect on the teaching of geography in Japan, especially through his textbooks. His philosophy was profoundly influenced by his membership, from the 1920s onwards, of the Orthodox Nichiren sect of Buddhism: he emphasized a system of 'value-creating education'. Some of his views were opposed by the World War II government of Japan, and he died in Sugamo Prison in Tokyo in 1944, partly as the result of malnutrition.

Thomas Robert Malthus (1766–1834) was one of the first teachers of the subject of political economy in Britain. Besides important work on rents and prices, he will be remembered for his *Essay on Population*, the first edition of which appeared in 1798. This sobering work, suggesting that human populations grew more rapidly than the rate of increase of means of sustaining them, had an important effect on the development of demography. Later editions of the *Essay* were rather more optimistic in outlook. His ideas influenced Charles Darwin (1809–1883) in his elaboration of his notion of evolution through natural selection, and have been advanced, criticized and re-advanced in discussions of population and population geography again and again over the last two centuries.

Akira Nakanome (1874–1959) is the second Japanese subject in this volume. He also was something of a polymath, being fluent in many languages (European and Asian). He studied in Europe under Penck and Brückner, was one of the first two professors of geography in Japanese higher education, and attempted to apply the ideas of his European teachers in East Asia. He contributed significantly to the development of 'homeland studies' or local geography in Japan.

John Ogilby (1600–1676) is the second subject from the seventeenth century whose life is revealed in this volume. In a fascinating but varied career he was by turns translator, poet, dancer, dancing teacher, theatre manager, and compiler of atlases and other geographical works. His *Britannia* (1675) is described as one of the most important works in chorography and cartography from the Restoration period. His standing as publisher and courtier placed him in an important position in relation to networks of influence and royal patronage through which much of science (including geography) was undertaken.

Thomas Pennant (1726–1798) occupied a similar important position in the networks of influence and diffusion of ideas a century or so later (although he was from Wales, rather than England or Scotland). Many readers will know his name as the receiver of many letters of Gilbert White (1720–1793) published in *The Natural History of Selborne* (1789), and he was indeed a prolific correspondent. He travelled widely within the British Isles and on the continent, and published a great deal on natural history and on the society and landscape of places he visited: some of his books could be considered very early tour guides.

Charles-Eugène Perron (1837–1909) was a Swiss cartographer who did much of the cartographic work for the great French publication *Nouvelle Géographie Universelle* of Elisée Reclus (1830–1909), and who also pioneered *cartographie nouvelle*, which involved producing maps through the photography of relief models to give an impression of height. Perron had an interesting and varied career, and this essay gives brief glimpses of life in Tsarist Russia and the activities of late nineteenth-century anarchist groups with which Perron was associated.

Pedro C. Sánchez Granados (1871–1956) provides an example of a genre important in the development of the geography of some Hispanic countries, the 'geographical engineer'. Trained in Mexico City in surveying and mining engineering, he later became an important figure in the fields of geodesy – the measurement of the earth – and cartography, taking part in important international co-operative projects with specialists from the United States, Canada and countries from Latin America. Later he taught geography at the National University of Mexico, where he shared in the diffusion of European determinist ideas into Mexican geography. He organized and founded the Pan-American Institute of Geography and History (PAIGH).

The final essay in this volume describes the life of Zheng He (1371–1433) from Yunnan – eunuch in the Chinese Emperor's court, soldier, navigator and trader – who in a series of seven voyages explored much of the Indian Ocean basin, bringing it into Asia's purview when that region had been barely glimpsed by European navigators. Some of these voyages were massive enterprises: one is said to have included 27,000 individuals, including diplomats, soldiers, traders, priests, technical personnel, translators and medical doctors. An undertaking indeed.

Patrick H. Armstrong
Nedlands, Western Australia

Geoffrey J. Martin
New Haven, Connecticut

John Adair

1660–1718

Charles W. J. Withers

John Adair was a mathematician, map-maker, natural historian and antiquarian. In his mapping work, Adair undertook both terrestrial and ship-based coastal surveys of Scotland with a view to improving coastal navigation and foreign trade and providing up-to-date county maps of Scotland. His use of circulated queries about Scotland in 1694 places him alongside contemporaries such as Sir Robert Sibbald for Scotland, John Ogilby (also in this volume) and Robert Plot for England and Sir William Petty for Ireland in elucidating useful natural knowledge from persons whose social status suggested them to be reliable commentators on the state of their nation. In the combination of his several interests and practices, Adair must be seen as typical of that *virtuoso* interest in knowing one's nation in all its respects that was so typical of geographical knowledge and of empirical science in the second half of the seventeenth century. Adair was certainly held in considerable esteem during his lifetime: he bore, at one time or another, the titles of King's Geographer, Hydrographer Royal, Her Majestie's Geographer, the Queen's Geographer and Geographer for Scotland. Yet, whilst his limited output shows him to have been a first-class map-maker, difficulties over funding, his own mismanagement of available finances and competition from others working to the same ends all combined to restrict his achievement as a whole. Even so, his mapping work in particular is crucial in revealing how the early modern state turned to maps in order to 'make space' for itself as a political and economic entity.

1. Education, Life and Work

John Adair was born on 2 September 1660, in the parish of South Leith, Scotland. Nothing is known either of his parents or of Adair's early education. The first firm reference to Adair as a map-maker dates from May 1681, when he was granted a licence by the Privy Council to survey Scotland and produce maps. This was prompted by an earlier request from Moses Pitt, who wanted maps of the Scottish counties to include in his *Great Atlas* of which, by 1681, two volumes of an intended twelve had already been published. Certainly, in May 1681, Adair both signalled his intentions and requested assistance of interested parties in an 'Advertisement anent surveying all of the Shires of Scotland and making new Mapps of it'. His first map, of Clackmannanshire, dates from 1681. By 1682, Adair had produced maps of the Roman Camp at Ardoch – the first expression of a long-run interest in Scotland's Roman antiquities – and been commissioned by the Geographer Royal, Sir Robert Sibbald, to undertake the maps for the latter's intended two-volume description of Scotland ancient and modern. Sibbald termed Adair 'mathematician and skilful mechanick'. In later years, the two were to be in dispute over contractual matters.

By an Act of Parliament of 14 June 1686, Adair's mapping projects were funded from an annual tonnage levy of one shilling (Scots) on all native ships over eight tons, and two shillings for foreign ships, to be paid annually for five years. The tonnage levy was to be an important, if alone insufficient, source of funding for Adair. Some time before September 1687, Adair married Jean Oliphant. No formal records exist either of the date or place of marriage or of the birth of known children: James (the eldest son), Patrick, who was a divinity student, or the daughters, Jean and Anna. At about this time, Adair began to get involved with the Royal Society in London. He presented a paper on the capercaillie, Scotland's largest native wild bird, and showed several of his maps at meetings of the Society in November 1687. He was made a Fellow on 30 November 1688.

By the early 1690s, Adair's work was being hindered by lack of money and by disputes with Sibbald. By August 1692, the Privy Council noted that he had completed ten sea maps, for which he had received £120 Scots from the tonnage levy although the work had cost him twice as much, and ten county maps, for which he had received less than £50. By August 1694, Adair's expenses sustained in producing maps of Scotland were nearly three times as much as he had received from the tonnage levy and other sources. The tonnage levy in support of his mapping was increased four-fold on 16 July 1695. In the summer of 1696, however, representations were made to the Privy Council from the Countess of Wemyss, amongst others, that the increased levy was proving prejudicial to the export of Scottish coal and salt on foreign ships. Her request that the imposition in favour of Adair be delayed was agreed, despite Adair's counter-claim that accurate maps were an incentive to foreign trade and that a shortfall in funds would restrict his work. In October 1696, the Privy Council reduced the tonnage levy for foreign ships to eight shillings per ton.

Adair's work was also hindered by another's claim to the tonnage levy as a source of support for work on the geography and history of Scotland. John Slezer, who had been from 1671 Chief Engineer with the army in Scotland, had published his pictorial account of Scotland, *Theatrum Scotiae*, in 1693. This work was likewise to have involved Sibbald, but Sibbald and Slezer also disagreed over contractual matters. Slezer intended a further work, 'Of Ancient Scotland, and its Ancient People'. The Scots Parliament, in approving Slezer's work, added Slezer's name to the beneficiaries of the 1695 Tonnage Act. Privy Council records for August 1697 show tonnage funds being given to Slezer and much smaller sums to Adair and record disputes between the two over funding for their respective works.

Adair was again involved with the wider scientific networks of London's Royal Society in 1697, being invited by Hans Sloane and Charles Preston to correspond over the geography and natural products of the north-west Highlands and Outer Isles. Adair set out for the Western Isles, via the Orkneys, in May 1698, on board the *Mary* of Leith, under the command of John Whyte. He was accompanied by Martin Martin, a Skye-born man then establishing a reputation for himself as a correspondent on behalf of the Royal Society. Adair's letter of 20 December 1698 to Hans Sloane makes clear that his intentions were to publish a description of the area, together with maps. Adair's plans received a setback, however, when Martin Martin published his own *A Late Voyage to St Kilda* in 1698. The expense of this voyage and other survey work probably also hindered the publication of Adair's work. A testament of 2 August 1699 from Thomas Whyte, ship's captain and son of the skipper of Adair's 1698 voyage, noted that Adair owed £480 for that earlier voyage.

Adair's *Description of the Sea Coast and Islands of Scotland*, which concentrates upon Scotland's east coast, was published in 1703. An intended second part was never published, although records from 1704 show further maps in an advanced state of readiness. A further tonnage imposition in support of Adair's work was ratified by Act of Parliament on 25 August 1704. In 1706 he was active in surveying the Shetland Isles. In 1707 he produced, in manuscript, a 'Short Account of the Kingdome of Scotland' with special reference to its coasts and fishing. From 1708 to 1712 Adair was active in plans for new docks at the port of Bo'ness, and he was active, too, in surveying the port facilities of south-west Scotland. A list dated 2 June 1713 enumerated nineteen maps not yet printed: most were of Scotland's west coast.

By late 1715, however, Adair's surveying activities were being curtailed by gout, although he was active in the summer of 1716 in preparing schemes, on behalf of the town's council, to control flooding in Perth. Adair died at home in the Canongate in Edinburgh on 15 May 1718. He had, in a disposition of 16 May 1698 in favour of his wife, sought to provide for his family upon his death, but records of July 1718 reveal considerable debts, the result, chiefly, of insufficient funds to support the outlay used in producing detailed maps of Scotland from field survey. From early 1719, his wife was actively seeking redress from debtors. She secured an annual pension of £40 in 1722, back-dated to Adair's death. Further sums were forthcoming from the Dukes of Queensberry and Argyll, yet she was also forced to sell many of Adair's instruments to make ends meet.

2. Scientific Ideas and Geographical Thought

Adair's importance principally rests in two areas: his 1694 *Queries*, and what they signify about the social context to the production of reliable natural knowledge, and, more importantly, his mapping and survey work. Both are of importance for what they reveal about the means to geographical knowledge in the later seventeenth century.

Adair's 1694 *Queries* – of which only two printed copies appear to have survived – was a list of fourteen points on which he sought information with a view to 'A True description of Scotland ... & Natural curiousities and antiquities'. In his use of circulated queries to members of Scotland's nobility, gentry, and, we must presume, ministers, Adair was doing the same as many other geographer-natural historians in that period. The use of circulated queries in this way may, for Britain, be considered the intellectual ancestor of the Census. In the context of their time, however, queries of this nature should more properly be seen both as the development of national geographical knowledge from people of a particular social status, and, importantly, part of the development of empirical natural philosophy. Yet while his use of such queries as a geographical method marks Adair as of his time, the nature of the questions asked suggests, in their emphasis on more strictly antiquarian and natural historical matters, an earlier tradition, different from the utilitarian enquiries of contemporaries such as John Ogilby or Sir Robert Sibbald. Adair asked, for example, about 'any strange Appearances in the air', 'If any odd insects have been observed', 'If the Rivers, lakes and rivulets have anything perculiar to them', 'If there be anything extraordinar about foor footed beasts', and whether men and women 'have been attended by any thing not common' during their lives. He was also keen to know about 'the Old Camps, Forts, Artificial Mounds, Cairns or heaps of stones up and down the Country, and what accounts or traditions are [held] about them'.

It is not appropriate to draw too firm a distinction between Adair's emphasis upon what contemporaries would have understood as 'the curious', even knowledge of 'the vulgar sort', and the more evidently utilitarian rhetoric of men like Ogilby or even Robert Boyle, for example. Their purposes differed and there was anyway a strong sense amongst natural philosophers at this time that *all* knowledge was useful knowledge. Yet it is hard to avoid seeing the content of Adair's queries as indicative of earlier antiquarian traditions and, in this sense, of Adair as something of a *virtuoso*. In his mapping, by contrast, Adair sought to provide up-to-date maps of Scotland through use of the latest instruments and methods. Such a conjunction of interests between, broadly, interests in the unusual and antiquarian and the practices of modern mapping was not unusual: to suppose otherwise would falsely simplify our understanding both of Adair and of his time.

It is likely that Adair was introduced to more senior figures in the world of Restoration survey and science whilst a young man; Robert Hooke's *Diary* for 12 January 1676 records a meeting between himself, John Ogilby and others at Garraway's coffee-house in London at which Adair was present. It has been argued that map-making and survey assumed a new social and political significance in the late seventeenth century. This is certainly supported in Adair's case. The licence granted to Adair in May 1681 by the Privy Council was motivated not just by Adair's invitation from Moses Pitt but by recognition that extant maps of Scotland, notably in Blaeu's 1654 *Atlas*, needed 'reformation'. It is interesting to note that Adair's 1681 map of the small county of Clackmannan was presented to the Council as proof of Adair's competence 'to make survey of the kingdome and prepare maps of the same'. Yet nothing is known of exactly when Adair began the work for that map.

Adair completed a map of West Lothian in 1681, and he had surveyed and mapped Midlothian and East Lothian by 1682. Not all were the result of survey. His 1682 map of Perthshire is described as a 'rude plan without surveying'. He completed, on the basis of survey, 'The Hydrographical Mappe of Forth' and 'The Mappe of Straithern, Stormont and Cars of Gourie' in 1683, and one of 'The East Part of Fife' in 1684. Others of the country around Stirling and the Firth of Clyde followed in 1685 and 1686. His work by this time was destined not for Pitt's intended atlas but for Sir Robert Sibbald's project for a two-volume atlas of Scotland, announced in his 1683 'An Account of the Scottish Atlas or the Description of Scotland, Ancient and Modern'. Adair was there charged to do the maps. Sibbald's patent as Geographer Royal effectively forbade anyone else from publishing maps except with his permission, as well as demanding that Adair channel his work in one direction. Adair was indebted to local merchants as early as 1682. It is this background of professional restriction and financial uncertainty, and not just the Act alone, that makes the 1686 Tonnage Act in favour of Adair important in understanding the development of his ideas and thought.

The 1686 Act is a splendid illustration of the early modern state's need for accurate maps. The Act speaks of how 'exact Geographical Descriptions of the several shires within this kingdom will be both Honourable and Useful to the Inhabitants; and the Hydrographical Description of the Sea-Coast, Isles, Creiks, Firths and Lochs, about the kingdom, are ... most necessary for Navigation, and may prevent several ship-wrecks. The want of such exact maps, having occasioned great losses in time past. And likewise thereby forraigners may be invited to trade with more security on our Coasts.' Adair's earlier work and his 'great skill, Diligence and Qualifications' were cited. Local men were to be appointed from each parish in support of his work.

Adair is to be understood in this context, then, as an agent of the Scottish state: the surveyor as civil servant. It may even have been with some sort of official warrant that he visited the Netherlands in 1687 to purchase mapping and survey instruments, returning in February 1688 with the engraver James Moxon. Yet it is clear, too, that Adair's plans were always frustrated by the failure of the state to fund his work adequately. It may have been for this reason as well as for the political patronage that it offered (and, possibly, for the status it afforded his patrons) that Adair undertook private commissions on garden design for men like Sir William Bruce of Craigiehall.

In his 1695 'Representation to HM High Commissioner anent the Surveying of the Kingdom and Navigating the coasts and islands thereof', Adair suggests that his work between 1681 and 1686 was, effectively, abandoned by him because of lack of funds, that the 1686 Act was crucial in affirming survey work and map production as 'honourable and useful' pursuits, and that he was still incurring considerable personal debts on behalf of his country: '... the buying copper plates, cutting of mapps, bringing home paper, and gathering in the accounts of naturall curiousities and monuments of antiquity through this kingdome for making up the descriptions was performed at the expence of the said John Adair'. This Representation prompted the further tonnage levy of 1696, monies from which paid for Adair's survey work on Scotland's west coast, chiefly around the Firth of Clyde, in 1696 and in 1697. Adair's travels to the Outer Hebrides in 1698 reflected a wider interest in the unknown geography of these islands. In that regard, Adair was working both as part of his larger plan to map Scotland and on behalf of the Royal Society in London. Full expression of his own work on the Outer Hebrides was certainly hindered by Martin Martin's publications: the fact that Sibbald in a later letter noted of Martin that Adair 'treated him scurvily' suggests that what was, in effect, a geographical and survey expedition to an almost unknown part of Scotland was marred by personal issues. Yet surviving manuscript maps from about this time of Adair's work in the Outer Isles reveal high standards of survey skill, standards apparent in his 1703 *Description*.

Adair had long-running interests in Scotland's natural history: his vast collection of Scottish shells was particularly commented upon by contemporaries. From the early years of the eighteenth century, he produced, in manuscript, several short works on Scottish fisheries, noting chiefly the benefits to the economy were ports to be developed and reliable maps of coastal waters provided. Surviving letters from him show that he petitioned members of Scotland's Parliament about the economic potential of fishing. His 1707 'Proposals for the Fishery in Scotland', written in a strictly utilitarian style, gave details of the men and equipment needed if Scotland were to take best advantage of this natural resource. Whilst such interest stemmed from his field observations, it is also likely that Adair was at this time influenced by discussions over Scotland's parliamentary union with England. The fact that the 1707 Union of Parliaments abolished both the Scottish Parliament and the Privy Council only made Adair's funding situation worse. In March 1714, Adair was described in customs accounts as 'Geographer for Scotland', but this was largely an honorific title. Illness by then meant he was failing to meet commissions – such as the maps for Alexander Pennecuik's *A Geographical and Historical Description of the Shire of Tweeddale*, published in 1715 – and his final years were again marked by debt.

3. Influence and Spread of Ideas

It is probably fair to describe Adair as an under-achiever, despite his extensive fieldwork, high standards, use of modern instruments and methods, and the production of a few high-quality maps. His several titles and the fact that he was made a Burgess of three Scottish towns during his lifetime suggests considerable contemporary esteem. Yet too much of his work was left in manuscript, and, in consequence, Adair's importance has been overlooked and his reputation neglected. Adair's failures were not all of his own making. Funding was never secure: the 1686 Act, for example, considered the levy on tonnage and support from towns only a voluntary not a statutory charge. Although the sum of £21,339 Scots was collected between August 1695 and August 1698, of which Adair received £12,840, this was insufficient to support him. Disputes with Sibbald, Slezer and Martin did not help his cause. Survey work was more time-consuming than other forms of geographical knowledge, and for that reason, too, he found his own plans for publication more than once thwarted by others. A collection of Adair's maps was handed over to the state in 1723 by his wife, but a fire in the Exchequer Office in 1811 is thought to have destroyed them.

There is no doubt that Adair was capable of excellent work. Yet, in the longer run, his importance rests more, perhaps, in what his life reveals of the complex practices of late seventeenth-century geographical knowledge and of the dangers inherent in undertaking too much with too little support, even from the very nation he so desired to delineate accurately.

Bibliography and Sources

1. SELECTED BIBLIOGRAPHY OF WORKS ABOUT JOHN ADAIR

Inglis, H.R.G., 'John Adair: an early map-maker and his work', *Scottish Geographical Magazine*, Vol. 34 (1918), 160–6.

Moir, D.R.G., *et al.*, 'John Adair', in D.R.G. Moir *et al.* (eds), *The Early Maps of Scotland*, Vol. 1, The Royal Scottish Geographical Society, Edinburgh, 1973, 65–78.

Moore, J.N., 'Scottish cartography in the later Stuart era, 1660–1714', *Scottish Tradition*, Vol. 14 (1986–7), 28–44.

Vasey, P., *John Adair: Geographer and Surveyor: A Source List*, Scottish Record Office, Edinburgh, 1998.

Adair manuscript material is fairly extensive, and is held chiefly in the National Archives of Scotland, and in the National Library, Edinburgh. The Vasey source list above gives a full chronological listing of what is available for scholars, although not a complete indication of the contents of all manuscript material. Some of

Adair's manuscript maps of the Outer Hebrides have recently (1997) been discovered in HM Admiralty Office in Taunton. There is no portrait of Adair; it is unlikely one was ever painted.

2. SELECTED BIBLIOGRAPHY OF WORKS BY JOHN ADAIR

A listing of all of Adair's intended publications and the several short works in manuscript is given in Vasey (1998). The interested reader is also directed towards Adair's 1704 'Account of the progress by John Adair as to the survey of the Kingdom' which contains lists of Adair's plans to date, including those in his 1703 *Description* (below). It also has his 'A Journal of a Voyage to the North and west Isles of Scotland in the year 1698' and lists of other sea maps 'surveyed and finished' and 'doing but not fully finished'; and 'land maps' 'surveyed and perfected' and 'doing but not fully finished'. This is printed in *The Miscellany of the Bannatyne Club*, Vol. 2, Edinburgh, 1850, 351–88.

1703   *Description of the Sea Coast and Islands of Scotland*, Clark, Edinburgh.

Chronology

1660	Born 2 September, South Leith parish, Edinburgh
1676	Met Ogilby and Hooke in Garraway's Coffee House, London
1681	Undertook and produced first map of Clackmannanshire. Granted licence by Privy Council to undertake mapping of Scotland
1683	Commissioned by Sir Robert Sibbald, Geographer Royal, to undertake maps for Sibbald's intended two-volume description of Scotland
1684	Sketches of Scottish birds included in Sibbald's *Scotia Illustrata*
1685	Made a Burgess of Stirling (and is noted in burgess rolls as 'Surveyor of this Kingdome')
1686	Mapping work funded by Tonnage Act, 14 September
1687	In Netherlands, purchasing survey instruments and contracting James Moxon as engraver. Married Jean Oliphant some time before September 1687
1688	Made a Fellow of Royal Society 30 November. Recorded in manuscript sources as 'King's Geographer'
1694	Printed and distributed list of fourteen *Queries* with a view to a 'True description of Scotland'
1695	Tonnage levy in support of his mapping increased four-fold, 16 July
1696	Tonnage levy on foreign ships in support of his mapping reduced. Surveying Scotland's west coast
1697	In dispute with John Slezer over support from tonnage levy
1698	Sailed to the Western Isles and to St Kilda, from May 1698, on behalf of the Royal Society, in company with Martin Martin. Made provision for wife after his death

1699	Made a Burgess of Canongate, Edinburgh
1701–3	Actively surveying in Northern Isles, working on Scotland's fishing industry
1703	Publication of *Description of the Sea Coast and Islands of Scotland*
1704	Further tonnage levy in support of his mapping, 25 August
1706	Made a Burgess of Aberdeen
1707	Produced, in manuscript, 'Short Account of the Kingdome of Scotland'
1708–12	Engaged in survey work and in planning new dock facilities in port of Bo'ness
1716	Working with Perth Town Council to review schemes aimed at flood prevention in the town
1718	Died 15 May in Edinburgh

Charles W. J. Withers is Professor of Geography at the University of Edinburgh.

James Cook, RN

1728–1779

Jill Rutherford and
Patrick H. Armstrong

Courtesy of the National Maritime Museum, Greenwich, UK

Born the son of a farm labourer, Cook was self-taught and self-assured; he became a navigator, explorer, hydrographical surveyor, astronomer and an enlightened leader of men for his times. He achieved much in his 50 years. His writings were an accurate and scientific record of the geography and anthropology of the lands he visited and his map-making skills were excellent. James Cook was a father-figure to his crews, from whom he received love and respect, but he had little time to be a family man. Sauerkraut (pickled cabbage), hygiene and a strict disciplinary regime kept his crews alive and healthy, and without these his long voyages of discovery would not have been possible. He received fame and credit in his lifetime but revealed little about his character apart from the outward signs of perseverance, resolution and courage.

1. Education, Life and Work

James Cook was born on 27 October 1728 in the village of Marton, Cleveland (North Yorkshire), England. He was the second son of James Cook (Senior) who was employed as a day labourer at the time of the younger James' birth. Much has been said of his lowly birth in a small two-roomed mud cottage. Of James' seven siblings little is known except that five of them are buried in Great Ayton churchyard, just a few miles from Marton; four died before reaching five years of age, and John, the eldest brother, at the age of 23 years. James Cook died aged 50, having started his first long voyage of discovery ten years earlier. His wife survived both him (by 56 years) and all their six children, seeing in the age of steamships.

In Marton, Cook attended school and learnt to read. When he was eight years old, his father gained a better position as hind or foreman to a Mr Skottowe, who paid for the boy to attend the day school where he received a grounding in literacy and mathematics. For a short time after leaving school, Cook worked as a stable boy on a farm but he left farming at the age of sixteen years to become apprenticed to Mr William Sanderson, a haberdasher in the coastal fishing village of Staithes some 10 miles (16 km) north-west of Whitby. In Staithes' small harbour was a fleet of Yorkshire cobbles. Talk would have been of the sea and ships, and it is hardly surprising that a future as a haberdasher did not beckon him.

After some eighteen months in Staithes, the incident of the South Sea shilling occurred. A customer paid for goods with a shiny South Sea shilling and Cook exchanged it for one of his own. Mr Sanderson missed it from the till and accused Cook of its theft. He later believed Cook's explanation but the incident was perhaps the impetus for change. Soon after, Cook went to Whitby, where he was bound apprentice for three years to Mr John Walker, a Quaker shipowner and master mariner. At this time, Whitby was in the midst of the greatest expansion in its history. Cook, aged eighteen, found himself in a lively and growing port. The alum industry, along the coast, required huge amounts of coal for roasting alum shale, and enough alum was made to satisfy the needs of the home market and export. Shipbuilding yards were springing up along the river banks. An Act of Parliament in 1702 imposed a charge of one farthing per chaldron (about 2.5 tonnes) on all coal shipped and goods moved. The revenue was spent on harbour improvements which led to the flourishing of new industries. When Cook arrived there were some 180 ships of 80 tons or more registered in Whitby. Cook's first ship was the *Freelove*, a collier of 450 tons, on which he made three voyages. He moved then to the *Three Brothers*.

During his apprenticeship, and during winters when the colliers were laid up, Cook lodged in the house of Mr Walker. Here he continued to study, possibly at night school, was noticed to be studious and serious and became respected by the family. Two further voyages 'before the mast' in the *Three Brothers* were to Norway, and then, in 1750, he moved to the *Mary*, owned by a John Wilkinson. A further move to the *Friendship* followed in 1752 and after five voyages as mate, he was offered her command.

Altogether, Cook spent nine formative years in North Sea trade. Even here, charts were unreliable; the approaches to the Thames were particularly difficult, with shifting sandbanks and currents. He gained a reputation for both daring and caution, knowing the best course almost by instinct. The *Friendship* was lying in the Thames in the summer of 1755 when Cook left it to join the Royal Navy. (It has been asserted that this surprising move might just possibly be because of some involvement, or alleged involvement, in smuggling.)

He started as seaman and within a month of joining, perhaps aided by a letter from John Walker, he was promoted to master's mate on HMS *Eagle*, a fourth rate 60-gun ship. Cook served for two years on the *Eagle* and saw action off the French coast. In 1757 he passed for master at Trinity House and joined HMS *Solebay*. As master, he was ship's navigator and would handle her in action, write up the log book and keep the rigging in good condition. On his twenty-ninth birthday he joined HMS *Pembroke* under Captain Simcoe which left England for Halifax, Nova Scotia, in April 1758.

Although a relatively short voyage, this was perhaps the longest so far in Cook's career. Twenty-nine men died of scurvy and the ship was too late to join the assault on Louisburg which commanded the entrance to the St Lawrence river. With Louisburg fallen, the St Lawrence was open for an attack on Quebec but the French had removed navigation markers and there were no charts. Cook spent the winter improving his knowledge of navigation and surveying. When the ice cleared in the spring of 1759, Cook on the *Pembroke*, in a squadron under Admiral Durrell, went ahead to clear some fortifications and chart the channel. By the end of June, the channel had been marked, the fleet sailed up river and Quebec was captured on 18 September 1759. Cook remained behind as master of the flagship *Northumberland* when the fleet left. He continued his surveying of the St Lawrence and Halifax and started on the Newfoundland coast. By then it was the end of 1762 and he returned to England after five years away. In December 1762 he married Miss Elizabeth Batts. The union appears to have been a happy one, lasting sixteen years until Cook's death, but taking into account Cook's long absences, they would have spent no more than four years together.

Because of his success in charting the St Lawrence, Cook was promoted to King's Surveyor in April 1763, and sent back to Newfoundland. He surveyed the St Pierre and Miquelon Islands, which were about to be returned to the French following the Treaties of Utrecht and Paris. After a winter in England, where he saw his son James for the first time, he was given his own ship, the schooner *Grenville* in which he was Master – his first Royal Naval command. The next four years followed the same pattern – survey of the coast and harbours of Newfoundland in the summers and return to England to prepare his charts each winter. In August 1766 there was an eclipse of the sun which he observed from Burgeo Island near Cape Ray. His measurements appeared in the Royal Society's *Philosophical Transactions* for 1767. The astronomer John Bevis claimed that he was 'a good mathematician and very expert in his business'.

Edmund Halley, in 1716, suggested that the distance from the Earth to the Sun could be calculated by timing the transit of Venus across the face of the Sun, but measurements were needed across the surface of the Earth for parallax observations. There were unsuccessful measurements in 1761 and it was calculated that in June 1769 another transit of Venus would occur. The Royal Society was laggardly in preparing for observations of this transit and it was a mere nineteen months before it was due that the King was the choice of petitioned and an expedition devised. Cook, aged 40, was named as the leader of the expedition as the result of a compromise between the Royal Society and the Admiralty. This was after some bickering, as Mr Alexander Dalrymple was the choice of Nevil Maskelyne (the Astronomer Royal). Dalrymple had caught the attention of Europe by suggesting that the land-mass of the Northern Hemisphere must be balanced by a similar mass in the south – *Terra Australis Incognita*, the fabled southern continent. (Cook followed Bougainville in being sceptical of its existence.) However, Dalrymple also insisted on being granted a naval officer's commission as commander of the expedition. This the Admiralty would not allow, and he withdrew.

The Royal Navy was anything but slow in finding a ship and after the Board suggested that a 'cat-built bark' would be suitable, the *Earl of Pembroke* (368 tons, and three years and nine months old) was purchased at Whitby for £2800. The cost of sheathing and fitting her for the voyage was £2294 and her name was changed to HMS *Endeavour Bark*. The *Endeavour Bark* was a typical East Coast collier ship. She was 106 feet (32.3 m) long and had an extreme breadth of 29 feet 3 inches (9 m) with a round bottom and shallow draught. Cook had now been promoted to First Lieutenant. Captain Samuel Wallis had by then returned from his voyage round the world in the *Dolphin* and reported that King George's Land (Tahiti) would be a good site for the observation. Cook rapidly recruited his crew and supervised the fitting-out of the ship. The hull was sheathed in wood (not copper, as Cook thought this would be hard to repair) and extra cabins and decks were created by a lateral division below deck. In this way, extra men were accommodated on board. Cook joined the ship on 7 August 1768 and carried with him sealed orders. The first part of these referred to the observation of the transit of Venus but the second was to remain sealed until at sea. These were:

> Whereas there may be reason to imagine that a Continent or land of great extent, may be found to the southward of the Tract lately made by Captain

> Wallis or the Tract of any former Navigators in Pursuits of the like kind ... You are to proceed to the southward in order to make discovery of the Continent above mentioned.

Joseph Banks (1742–1820), then a 25-year-old Fellow of the Royal Society, and Dr Carl Solander (1733–1782), as naturalists, joined the ship at Plymouth. Charles Green, an assistant at the Greenwich Observatory, and Sydney Parkinson, a natural history artist, were also part of the scientific entourage; the total on board was 94. Many of the officers were to survive this voyage and sail with Cook again. Most were young – Charles Clerke, master's mate, being one of the oldest at 25. Practically all enjoyed a drink. The crew of about 75 men were a mixed crowd of Britons, Irish, a Venetian and a Brazilian. Some were impressed men and all anticipated a harsh regime and cramped living conditions, ameliorated only by the rum ration of one pint 94° proof per day and Cook's paternal care. The ship left Plymouth on 26 August 1768.

Although it had just become available, Cook did not carry Harrison's chronometer on this first voyage. However, he did insist on antiscorbutics. Scurvy, caused by a deficiency of vitamin C in the diet, had long been recognized as a killer of sailors; it struck after only six weeks on salt rations. Cook insisted to the point of lashes that the men eat the antiscorbutics.

En route to Otaheite (Tahiti), the *Endeavour* stopped at Funchal (Madeira) and Rio de Janeiro. After three unsuccessful attempts the ship passed through the Strait of Le Maire, then passed Cape Horn and into the Pacific on 27 January 1769. Banks and Solander took every opportunity to go ashore and collect plants and animal specimens and observe the natives. As Cook later wrote to John Walker (his former employer): 'We arrived at Georges Island on the 13th April 69, having in our track thither discovered several islands which are of no great note'. Cook noted no ocean currents indicative of a large land-mass during the passage.

Third Lieutenant John Gore had been with Wallis in the *Dolphin* when he visited Tahiti two years previously. He later noted a degradation in the quality of life for the natives since their first visit.

> Almost everything was altered for the worse ... [There had been] a number of fine houses dispersed among the trees, many inhabitants of the better sort, a large number of canoes ... instead of all this found a few temporary huts with a few of the inferior sort of inhabitants.

The *Endeavour* spent three months in Matavai Bay and in that time Banks gained a good grasp of the language. Most of the officers and gentlemen kept journals, as they were no doubt aware of the significance of their visit and the effect that such visits were having on the natives. The crew built a sturdy wooden fort at Point Venus for observation of the transit. The observation, according to Cook:

> proved as favourable to our purposes as we could wish. Not a Clowd (*sic*) to be seen the whole day and the Air was perfectly clear ... We very distinctly saw the atmosphere or dusky shade round the body of the planet ... Dr Solander observed as well as Mr Green and myself.

There were a few incidents between natives and men, which Cook dealt with fairly. Most of the crew took temporary wives but venereal disease (from previous visitors) or yaws (endemic to the Pacific) appeared amongst some 33 of the *Endeavour*'s men.

It was time to depart. Fort Venus was dismantled, Banks planted seeds of melons, limes, oranges and lemons and Cook was besieged by requests from young native men to take them too. He eventually took Tupia, a native chief and priest, and his boy servant. The *Endeavour* circumnavigated Tahiti and mapped 75 other islands in the group, which Cook named the Society Islands, and of which he took possession for the King.

Cook took his ship south, as instructed, but having not found 'the least visible signs of land' and as the weather was very bad with a continuous swell from the south, he turned west. On 7 October, land was sighted. This was the east coast of New Zealand. Again, in Cook's words to John Walker:

> a very small part of the West coast of which was first discovered (by) Tasman in 1642, but he never once set foot on it: this country was thought to be part of the Southern Continent, but I found it to be two Large Islands, both of which I circumnavigated in the space of Six Months ... [They] are together nearly as big as great Britain.

The natives were deemed not friendly by Tupia, who could converse with them. The weather was bad. Cook wrote in his log:

> Monday, 1 January 1770: ... but it will hardly be credited that in the midst of summer and in the Latitude of 35 degrees, such a gale of wind as we have had could have happen'd, which for its strength and continuance was such as I hardly was ever in before. Fortunately at this time we were at a good distance from land otherwise it might have proved fatal to us.

By mid-March, the *Endeavour* had rounded the southernmost point of South Island and the doubters recognized Cape Turnagain, again. Banks and the others at last conceded that this was not the Southern Continent. At this time, Banks's request to enter a narrow fjord was declined by Cook, who saw the danger of being in a west-facing fjord with a west wind blowing. Banks did not forget this event and recalled it 30 years later with displeasure.

Cook returned to a safe harbour in Cook Strait. His mission completed, he was to return to England. The historian Beaglehole identifies the four possible courses that he might have taken:

- sail west and around the Cape of Good Hope. This was well charted already and would not allow for further discovery.
- travel east across the southern Pacific and around Cape Horn. This was Cook's wish but the late season and state of the ship would have made this too dangerous.
- make directly for the East Indies in order to refurbish and outfit for the return.
- sail westward until reaching the New Holland coast, then turn north to reach the East Indies.

The officers were unanimously for the last option. The ship was now on the return leg to England. Cook

> steer'd for New Holland all the East part of which remain'd undiscover'd, my design being to fall in with the Southern part call'd Van Diemens Land, but the winds forced me to the northward of it about 40 leagues, so that we fell in

with the Land in the Latitude of 38° South.

Cook thus made no claim to have discovered Australia, being already aware of its existence. Quiros, the Portuguese navigator, reportedly coined the name Australia, and the Dutch, Chinese and Malays knew of it. The landfall, named Point Hicks after the lieutenant who first saw it, was too far north for Cook to identify that Van Diemen's Land was indeed an island (Tasmania) but the doubtful line on his chart suggesting it might be was verified by Bass, who sailed through Bass Strait. The first anchorage was a shallow bay, at first named Sting Ray Bay but changed to Botany Bay after Solander and Banks made so many discoveries of new plant species there. This place was to be the destination for Captain Arthur Phillip's first British settlement in 1788. Cook continued northwards, mapping, naming and gaining fresh supplies wherever possible.

In the sail northwards, the *Endeavour* was all the time being funnelled into the ever-narrowing channel between the Great Barrier Reef and the mainland (Queensland). Eventually, despite Cook's seamanship and navigating skills, *Endeavour* grounded on the coral and stuck fast. Even with 50 tons (including six guns) cast overboard, she remained stuck until hove off and then fothered with a sail (fother = stop a leak by covering it with a sail). It took two months to repair the ship beached on the mainland and they then sailed off the coast to the north end of the 'Labyrinth' before returning to map the coast from 13°S to the tip of Cape York. On 22 August 1770 Cook went ashore to Possession Island and claimed the lands for the King. They passed through Torres Strait:

> Thursday, 23 August 1770: ... the wind had got to SW and altho it blowed but very faint it was accompanied with a swell from the same quarter; this together with other concurring circumstances left me no room to doubt but we were got to the Westward of Carpentaria or the Northern extremely of New-Holland and had now an open Sea to the westward, which gave me no small satisfaction not only because of the dangers and fatigues of the Voyage was drawing near to an end, but by being able to prove that New-Holland and New-Guinea are two Separate Lands or Islands, which untill this day hath been a doubtfull point with Geographers.

The *Endeavour* headed for the New Guinea coast and then along to Batavia (Jakarta). They were laid up for ten weeks here in a port full of dysentery and malaria. Until this point on the voyage not one man had died from scurvy: three had drowned; Buchan, the draughtsman, had died of an epileptic fit in Tahiti; able seaman Sutherland of tuberculosis; boatswain's mate Reading of excessive rum; and the two servants of Banks of cold near Cape Horn.

Ironically, at Batavia, Surgeon Monkhouse was the first to die, then Tupia, followed by his servant boy. There remained fewer than twenty fit men to sail the ship and Cook recruited another nineteen hands to get the *Endeavour* home. But his comment on the grounding on the Reef, 'Here begin all our troubles', continued to hold true. On the way to Cape Town the deaths continued: Mr Green the observer, the artist Sydney Parkinson, Midshipman Monkhouse, the surgeon's brother, the one-armed cook and others, in all 34 before reaching Cape Town and another five at Cape Town or on the way home. One was Zachary Hicks, Cook's second lieutenant, who finally gave in to consumption. It must have been a heavy blow to Cook when his right-hand man died. The *Endeavour* anchored in the Downs on 13 July 1771.

As Beaglehole says, Cook did not discover *Terra Australis Incognita*, Tahiti, New

Zealand, Australia, nor the Torres Strait. But he did reinforce the belief that the southern continent was a myth, made possible creditable observation of the natives of Tahiti, mapped New Zealand in great detail, charted the east coast of Australia and confirmed the passage between Australia and New Guinea.

Within a month of his return, Cook was promoted to Captain and committed to lead another long voyage. This time Banks saw himself as the leader of the expedition to find or disprove the existence of *Terra Australis Incognita*, the Great Southern Continent. But he could not do it without Cook. Cook found the ships for this voyage: again, North-East colliers, originally named *Drake* and *Raleigh* but renamed *Resolution* and *Adventure* to avoid upset to the Spaniards. While these were undergoing refit, Banks commandeered *Resolution*'s great cabin and others for himself and his entourage: fifteen in all. He then had built another deck and raised poop. The first sea tests showed this to be unstable and unsafe and the ship was ordered to Sheerness, where the extra weight was removed. Failing to get his way with the navy despite wrathful letters, Banks withdrew from the expedition and went to Iceland instead. The Forsters, father and son (Johann Reinhold, 1729–1798, and Georg, 1754–1794), joined as naturalists and Cook finally took the Kendall chronometer, a replica of Harrison's H-4, and also three inferior timepieces which gave him the time at Greenwich wherever he went: he was thus able to calculate longitude accurately. The Kendall chronometer kept time to within a tenth of a second a day, which translates to within half a degree on a voyage from England to the West Indies. While a known sceptic on the southern continent, Cook was willing to go for further discoveries in the South Pacific. He had decided to sail by the eastern route via Cape Town and make two bases: Tahiti and Queen Charlotte's Sound. Tobias Furneaux was commander of the *Adventure* and while a good seaman, he was not of Cook's calibre. On this second voyage Cook lost four men, only one of these to sickness and not one to scurvy. Cook was obsessional about cleanliness on board and kept the men busy cleaning, particularly below decks. Furneaux's ship lost thirteen men, the cook from scurvy, and many more suffered from it but recovered. Dutch East Indiamen often lost half a crew to scurvy on long voyages. Conditions were nevertheless harsh: there were 118 on board the *Resolution*, which was 111 feet (34 m) long and 35 feet (10.6 m) wide and just 462 tons. There were rations for two years, including 1400 gallons (6400 litres) of spirits and 60,000 pounds (27 tonnes) of ship's biscuits and 20,000 pounds (9 tonnes) of sauerkraut, salted cabbage and portable broth, and 30 gallons (137 litres) of carrot marmalade.

For three Antarctic summers they sailed close to the ice barrier, reaching 71°S. In the southern winters they went north, adding the Cook Islands, the Friendly Islands and New Caledonia to the maps. On 23 March 1773, when Cook saw that the crew were at the end of their endurance, he made for Dusky Bay, New Zealand, where they rested for six weeks. The *Adventure* had become separated from the *Resolution* in fog but when the ships reunited in Queen Charlotte's Sound, Cook was not pleased with the state of his sister ship. Furneaux had on board the same antiscorbutics as Cook, but they had not been adequately distributed. The signs were that the *Adventure* was ready for a winter resting in the Sound. Cook then ordered that both ships would sweep the ocean between 41 and 46°S. The weather was bad with irregular swells, gales and torrential rain. Even worse, scurvy had broken out again on the *Adventure* and Cook was seething at the inability of his fellow commander to control it. They found Tahiti again by grounding on the reef but, once again, the island was much changed. Civil war had broken out again and Otoo was now chief. The crews recovered their health on the island and the *Adventure* gained another man – Omai, a quick-witted native who begged to be allowed to go with Furneaux. After leaving Tahiti on 18 September and discovering the Cook (Hervey) Islands and visiting the Tongan Islands they headed for New Zealand once more.

Yet again the ships lost each other in Cook Strait and this time Cook left instructions to Furneaux and gave him no rendezvous. Furneaux was free to return home, and this he did, via Cape Horn. This was perhaps a relief to them both. Cook continued to probe the ice pack for signs of a land-mass but was halted by field ice. Now satisfied after four months that there was no land to the south, they turned, found Easter Island and the Marquesas, and returned to Tahiti, where they found conditions to be much improved. They sailed on to discover many of the Tuamotu Islands, Society Islands, Tonga, and Fiji Islands until reaching what Cook called the New Hebrides (Vanuatu). From here they sailed south and touched at New Caledonia before returning to Queen Charlotte's Sound. Here the message left by Cook for Furneaux had been removed but none left in its place. The Maoris were uneasy and explained that a group had killed and eaten some white men. Cook suspected these were from the crew of the *Adventure* but this was not confirmed until they reached the Cape of Good Hope. On leaving New Zealand, Cook steered for Cape Horn, rounded it and then went across to South Georgia and the South Sandwich Group. Here neither Cook nor Clerke was initially sure whether these southern islands were isles or the point of a continent, but eventually Cook wrote in summary:

> I had now made the circuit of the Southern Ocean in a high latitude traversing in such a manner as to leave not the least room for the possibility of there being a continent.

At the Cape, they heard that the *Adventure* had been there twelve months before and one of the crew had indeed been eaten by Maoris. The *Resolution* returned to Plymouth in July 1775. This second voyage had lasted three years and eighteen days.

Naturally, Cook was fêted on his return, had an interview with the King and was made post-captain, Fellow of the Royal Society and Captain in Greenwich Hospital, a lucrative sinecure. In view of the inaccuracies resulting from the publication of John Hawkesworth's account of the first voyage, Cook was allowed to publish this second voyage himself. Omai, brought to England by Furneaux, was still a novelty doing the rounds of country houses and under the wing of Banks.

Even before the *Resolution* returned, another voyage of discovery was mentioned. This time it was to seek a North-West Passage from the Atlantic to the Pacific, so avoiding the two capes in reaching the East. John Cabot had first looked for this in 1497, finding Newfoundland but convinced it was Cathay. The Spaniards and Portuguese had effectively split the southern Atlantic between them, making it difficult for other nations to reach the riches of the East. The lure of a North-West Passage was strong and the British government saw the return of Omai to his land as the best cover story for an expedition from the Pacific side. Cook was to be advisor but could not hold back from another adventure. On dining with Lord Sandwich, Sir Hugh Palliser (Comptroller of the Navy) and other dignitaries, he offered his services and was met by a loud cheer. This time Cook had not the time to be as diligent in supervising the refit as in the past. The navy yard at Deptford was lax in the extreme and the choice of *Resolution*, although already well tested, was to turn out to be the wrong one. Despite a refit, her seams needed constant recaulking and her masts and spars required renewal. The *Discovery*, of only 295 tonnes, was the second ship. By the end of May 1776, the crews were complete. Cook's first lieutenant was John Gore and *Discovery*'s commander Charles Clerke. Among the officers were George Vancouver and a 21-year-old William Bligh.

Cook set out from England and was caught up by Clerke in *Discovery* after he had escaped from debtors' prison but contracted tuberculosis. Already the *Resolution* was letting in water but both ships made landfalls at Tasmania for mast repairs, as well as New Zealand. First Omai was taken home but his extravagant gifts and unwise disposal of these made his homecoming look perilous. They then discovered the Sandwich Islands, the largest of which was Hawaii, and stayed five weeks, finding the natives traded fairly and were of pleasant disposition. They then started to chart the North American coast. There were more repairs at King George's (Nootka) Sound on what later became known as Vancouver Island. A continued passage through the Bering Strait was blocked by ice to the north. At this point Cook ordered an attempted passage west, north of Siberia. By then it was September and the pack ice was advancing. Although the Siberian coast was sighted, they turned back. This was the first uncharacteristic, somewhat alarming, order that Cook had given; it came on top of his insistence that the men eat either the virtually inedible walrus meat or only ship's biscuit. At the end of August the ships retired for the winter back to Hawaii, where their reception was unexpected. The natives thought Cook to be the god Orono. The ships proceeded around the island purchasing provisions and all the time their reception was ecstatic. Even here though, Cook made two strange decisions. One was to allow women on board ship, hitherto forbidden, and the other was to cut the rum ration first to alternate days and then completely. This brought the crews nearer to mutiny than ever before. Gales that forced the ships away from land and split the *Resolution*'s sails did not help matters, but eventually they found anchorage in Kealakekua Bay. There Cook was greeted as a god and processed to a morai. In all their actions, the ships had fulfilled the legend of the god Orono and Cook was seen to be he.

The visit dragged on for too long and eventually the natives made it clear that their lavish hospitality was not infinite and they expected the god to depart. The leaving of the bay was slow and emotional, but once outside the wind got up without warning and the two ships had to weather the storm. *Resolution*'s foremast split and an old leak opened. They were forced to return, but the welcome had evaporated. Relationships worsened as theft was punished by floggings; the large cutter was stolen from the *Discovery*. Cook followed his usual practice and decided to take a hostage, this time the King, Terreeoboo. He went ashore with one officer and ten marines. On reaching the King, who agreed to come, Cook and his party retreated to the shore and the boats but a large crowd of angry natives gathered. The marines fired a volley, leaving nothing in reserve – it taking twenty seconds to reload – and turned and scrambled for the boats so tantalizingly close. Cook now appeared, the only sailor ashore, among the rocks. He was walking slowly, a hand behind his head as protection against stones. Around him were marines, warriors' corpses and other warriors hesitating to attack. Then one hit Cook on the back with a club and he staggered and fell on one knee, another drew a knife and stabbed him in the neck; he fell but tried to rise from the water several times until a frenzy of warriors attacked his corpse.

Four marines were dead as well as Cook. Over the following days parts of Cook's body were returned to the ships: his hands, scalp, skull and long bones were committed to the deep on 22 February 1779. Clerke took command and tried again for a passage through the Bering Strait, but was defeated by ice once more. A month later he was dead and Gore took command, with King in *Discovery*. Bligh navigated the ships home via Petropavlovsk, Macao and the Cape of Good Hope. They reached the Thames in October 1780 after four years and three months away.

2. Scientific Ideas and Geographical Thought

Since the time of Ptolemy (fl. AD 127–150), who may have gained the notion from Hipparchus, there were those who maintained that the Indian Ocean was bounded to the south by *Terra Australis Incognita* (the unknown south land). The Arabs, who traded around the Indian Ocean basin for many centuries before the advent of European explorers, knew the ocean to be open to the south, but accounts from that quarter had little impact in Europe. The 'Great South Land' concept gained further currency from the idea that the mass of land in the northern hemisphere must be balanced by an equivalent mass in the south. The idea was remarkably persistent; Vasco da Gama's (*c*. 1469–1525) rounding of the Cape of Good Hope in 1497 failed to dislodge it – the land-mass just retreated to the south. Magellan (*c*. 1480–1521; this series, Vol. 18), sailing through the strait that now bears his name late in 1520, believed that the land to the south of him, Tierra del Fuego, might be part of a larger land-mass. Occasional glimpses of islands and even icebergs in the two centuries that followed allowed the idea to persist even after Francis Drake demonstrated the island nature of Tierra del Fuego. Some maps showed Australia linked to the southern continent. A French chart of 1754 showed a large south polar land-mass extending north to incorporate New Zealand. The myth was from time to time fuelled by those whose tales were almost pure invention, such as Yves-Joseph Kerguélen who, in early 1772, reached what are now called the Isles Kerguélen, later describing them in glowing terms: '... intersected by woods and greenery ... inhabited and carefully cultivated'.

James Cook, therefore, following the secret instructions, after the observation of the transit of Venus, proceeded south, as far as 40° 22'S. There were not, as Cook himself put it, 'the least Visible signs of land'. Only then, when he was certain that there was 'no prospect of meeting with land', did he head towards the north and then westwards towards New Zealand. (His instructions were, should he find the southern continent, to explore and appraise it fully, and to claim it for the Crown of Great Britain.) On his second voyage he circumnavigated the earth in high latitudes. He crossed the Antarctic Circle on three occasions, the first on 17 January 1773. He was convinced that his was 'undoubtedly the first and only ship that ever cross'd that line'. He made another venture south some months later, and was again blocked by bad weather and ice. On 30 January 1774 he reached 71° 10'S, his furthest south, encountering much ice. He wrote: 'this Southern Continent (supposing there is one) must lay within the Polar Circle where the Sea is so pestered with ice that the land is thereby inaccessible.' The myth of a massive, possibly fertile, Great South Land had been dissipated once and for all.

Even more important was Cook's documentation of the lands that did exist. Before his day the Pacific was little known. Crossings had been infrequent, and its boundaries uncertain. Cook charted accurately the east coast of Australia, and much of New Zealand. On his last voyage he explored the west coast of Canada in one of the many government-sponsored, fruitless searches for the 'North-West Passage' that it was hoped would provide an alternative route from Europe to India. During this search he passed through the Bering Strait, and into the Arctic Ocean. He fixed the position of islands in the Pacific, and also in the southern Indian and South Atlantic Oceans, whose position had previously been in doubt. He produced charts of many of them. J.C. Beaglehole put it aptly when he said that the map of the Pacific 'was Cook's ample panegyric'.

But there was more to Cook than the navigator and hydrographer. The first voyage, the transit of Venus expedition, was one of the earliest where a corps of trained scientists had accompanied a naval expedition. As well as an astronomer and the two botanists Joseph Banks and Daniel Solander, there were natural history draughtsmen. The expedition was well equipped with scientific instruments. The tradition continued when the two Forsters were included on the second voyage. Thousands of natural history specimens, particularly plants, were collected. Cook demonstrated himself a splendid observer and no mean anthropologist, and his accounts, for example, of the customs, speech and language, dances and way of life of the Tahitians are most detailed. The pattern was established in which the aspirations of science mingled with imperial ambition.

But it was, as much as anything, for his improvement of the diet of ships' crews that James Cook received his Fellowship of the Royal Society (1772). He understood the importance of fresh foods, and where on long voyages these were impossible to provide, he substituted sauerkraut and malt. He encouraged the seamen to consume the sauerkraut by conspicuously providing it for the officers, telling the lower ranks that they could eat it if they wished. They did, and Cook lost remarkably few men to scurvy, the bane of many a sea-captain of his day. Thus long before the scientific nature of vitamins was comprehended, Cook virtually eliminated scurvy from his ships through diet. The consequences of this for the future of maritime exploration (and imperial development) were considerable.

Despite the manner in which Captain James Cook was eulogized, during his lifetime, at the time of his death, and thereafter, it must be remembered that he was a product of his own generation, and saw the world through the lens of his own time. Although lenient compared with many Royal Navy commanding officers of his day, he occasionally found it necessary to flog his seamen. He took hostages against the return of stolen equipment in Tahiti, and his treatment of some of the indigenous peoples with whom he came in contact was sometimes brutal. (It has been suggested that on his third voyage his own health may have deteriorated, impairing his judgement.)

He thought some of the dances of the young women 'indecent', using the same word to describe the manner in which young men and young women of those islands cohabited, sometimes actually copulating in public. In his ethnographic accounts he concentrated, as did many eighteenth- and nineteenth-century anthropological accounts, on the exotic and the bizarre. He considered it unlikely that many would venture south of the Antarctic Circle beyond the point that he reached: there was a touch of arrogance to him. He would be amazed by the enormous development of Antarctic research of the twentieth century, with literally thousands of scientists visiting the continent each summer season.

3. Influence and Spread of Ideas

From the time of his death, and also probably partly because of the manner of it, Captain James Cook, RN, FRS, was seen as the archetypical 'great explorer'. Responsible for discovering (and claiming) new lands, resolute to the point of obstinacy, professional in both his seamanship and hydrographic survey, concerned for the welfare of his men, Cook was considered something of a paragon, and an exemplar of his genre throughout the British imperial age. As such he served as an inspiration to dozens of explorers, navigators and hydrographers for two centuries – a tradition that included, for example, Robert FitzRoy, John Lort Stokes and Francis Beaufort (this series, Vols 11,18 and 19 respectively).

Cook's influence upon geographical thought – direct and indirect – should not be underestimated. Besides demonstrating conclusively that *Terra Australis Incognita* did not exist in the form envisaged by the ancient geographers, he delimited the boundaries of the lands bounding the Pacific Ocean, at the same time hinting at the presence and nature of Antarctica. The information he brought back from each voyage was eagerly awaited, and often widely circulated in books and journals, having an influence for many decades; for example Conrad Malte-Brun, a Dane who worked in Paris writing a *Compendium of Geography* between 1810 and 1829, included information on wind systems from the voyage. It should not be forgotten that the 'Transit' expedition was the first voyage of its sort on which scientifically trained people had been included. The botanical work of Banks and Solander was of far-reaching importance in the establishment of that science, not least because Sir Joseph Banks, on his return from his voyaging with Cook, held the post of President of the Royal Society for 41 years. Similarly, on the second voyage, Cook had with him the two Forsters, who collected many thousands of plant specimens. Georg Forster pointed out the patterning of climates on the eastern and western sides of continents, and was an important influence on the young Alexander von Humboldt (1769–1859), traveller and pioneer geographer. And it was Humboldt's *Personal Narrative of a Voyage to a New Continent* that triggered Charles Darwin's (1809–1882; this series, Vol. 9) desire to travel to see something of the world and to make a contribution to science. The intellectual link between the two circumnavigators, Cook in the eighteenth century and Darwin in the nineteenth, is indeed a strong one. (Darwin was, incidentally, familiar with the writings of Captain Cook; there was a copy of Cook's *Voyages* aboard the *Beagle*, and Darwin quotes Cook in the *Voyage of the Beagle*.)

Captain James Cook, RN, has been something of an icon in Australasia, and his portrait has appeared on stamps and coins. A measure of the affection in which his memory continues to be held is the fact that 'Captain Cook's Cottage', moved from Yorkshire to Fitzroy Gardens in Melbourne in 1934, and now surrounded by a charming English-style garden, is one of the most popular tourist sites in Australia. Since its restoration in 1978 – the 250th anniversary of Cook's birth – it has attracted approaching two million visitors. (In fact the cottage was the home of Cook's parents, and evidence that the young James ever stayed there for more than the occasional visit is slim.)

The name of his bark *Endeavour* was remembered in the name *Young Endeavour*, a sail-training ship presented by Britain to Australia on the occasion of Australia's Bicentennial in 1988. An *Endeavour* replica built in Fremantle, Western Australia (1987–1994), and subsequently sailed to many ports in Australia and New Zealand and then to Europe (including Whitby) and North America, similarly perpetuates the image of Cook the navigator around the world. *Endeavour* has also been used as the name of a scientific journal.

Cook's name is recalled by dozens of place names in many parts of the world, but particularly in Australia, New Zealand and the Pacific. These include the Cook Islands in the South Pacific; Mount Cook, North Island, New Zealand; Cook River, South Island, New Zealand; Cook Strait, between the two islands of New Zealand; Cooktown, Queensland. However, it has to be said that with a rise in Aboriginal and Maori consciousness, the cult of the 'Great European Explorer' has declined somewhat in the Antipodes in recent years – particularly since 1988, the year of Australia's Bicentennial. Cook and his ilk may well have revealed the Pacific and Australasia to Europe, but there were others who had been seeking to understand Australian, New Zealand and Pacific island environments in their own ways for many centuries previously. Powerful though Cook's influence was, it is being increasingly realized that there were and are other geographies besides the

Eurocentric one. The deleterious effects on the society and environment of the indigenous peoples of the Pacific caused by contact with Europeans (and Americans) is also a matter that has been much discussed by scholars over recent decades.

Bibliography and Sources

1. ARCHIVAL SOURCES

Because of his association with Australia, it is unsurprising that a number of archival sources relating to Cook are now situated in that country. A number of Cook papers spanning the period 1762–79 are held by the Manuscript Collection of the National Library of Australia in Canberra. These include the *Endeavour* journal. The same library holds James Burney's journal of Cook's second voyage to the Pacific. The library also contains some papers of the Cook family (1776–1926). Cook's journals of the second and third voyages are in the British Library, as are many other manuscript maps and other papers relating to the voyages. A number of papers relating to Cook's life, work and the surrounding circumstances, together with related memorabilia, are held in the Captain Cook Museum, Whitby, Yorkshire. Other relevant documents are in the Mitchell Library, Sydney, New South Wales; the Public Record Office at Kew; and the Hydrographic Office at Taunton, Somerset. The Whitby Museum possesses the Muster Rolls, which are unique records of ships and their crews sailing out of the port of Whitby; each man on every ship's voyage also paid a contribution to a fund described as 'a New Year's gift to the distressed seamen and to the relief of their widows and education of their children in the Towne of Whitby'. Cook is listed in these records, which were kept from 1747.

2. SELECTED WORKS ON CAPTAIN JAMES COOK

Beaglehole, J.C., *The Life of James Cook*, The Hakluyt Society, Cambridge and London, 1974. A detailed biography; published as an accompaniment to the *Journals* (1955–67) below.

Cobbe, H., *Cook's Voyages and Peoples of the Pacific*, British Museum, London, 1979.

Fisher, R. and Johnston, H.J.M. (eds), *Captain Cook and His Times*, Croom Helm, London, and Douglas & McIntyre, Vancouver, 1979. An important series of essays, based on the proceedings of a conference, with several papers on the influence and reputation of the navigator, some of which aim to 'demythologize' the figure of Cook.

Frost, A., *The Voyage of the Endeavour: Captain Cook and the Discovery of the Pacific*, Allen and Unwin, London, 1998.

Hough, R., *The Murder of Captain Cook*, Macmillan, London, 1979.

Hough, R., *Captain James Cook*, Hodder & Stoughton, London, 1994.

Kippis, A., *A Narrative of the Voyages round the world performed by Captain James Cook, with an account of his life*, London, 1788. (Several subsequent editions.) The first biography.

Kitson, A., *Captain James Cook, R.N., F.R.S.*, John Murray, London, 1907.

MacLean, A., *Captain Cook*, Collins, London, 1972. (Later appeared in Fontana paperback.) A popular biography by a popular writer.

Moorehead, A., *The Fatal Impact: An Account of the Invasion of the South Pacific 1769–1840*. Hamish Hamilton, London, 1966.

3. CAPTAIN COOK'S WRITINGS

Bibliographic Note

It was the custom in the eighteenth century for the Admiralty to insist that all logs and journals kept on a voyage were passed over so that the authorized account of a major voyage could be published first. Nevertheless unauthorized diaries were kept, and (often very profitably) published soon after the voyage ended. The 'official' account of the *Endeavour* voyage, prepared by John Hawkesworth, was a melding of the journals of Banks and Cook, in places rather freely edited. It appeared, in a set of three volumes describing the explorations of several other navigators as well as those of Cook, in 1773. Cook's own journal was not published until 1893, and a facsimile of the original handwritten manuscript in 1955 (by the Hakluyt Society). By the time his second voyage was complete, his reputation was such that he was able to have a much greater influence over the editing. The editions listed below are regarded as definitive; several were used by the authors of this essay.

Andrew, D., *The Charts and Coastal Views of Captain Cook's Voyages*, 3 volumes, The Hakluyt Society, Extra Series No. 43, London, 1988.

Beaglehole, J.C., *The Journals of Captain James Cook and His Voyages of Discovery*, 3 volumes, The Hakluyt Society, Cambridge and London, 1955, 1961, 1967.

Douglas, J., *A Voyage to the Pacific Ocean in the years 1776, 1777, 1778, 1779 and 1780, ... Vols I and II written by Captain J. Cook, Vol. III by Captain J. King*, 3 volumes, London, 1784.

Hawkesworth, J., *An account of the voyages undertaken ... for making discoveries in the Southern Hemisphere ... performed by Commodore Byron, Captain Wallis, Captain Carteret, and Captain Cook in the Dolphin, the Swallow and the Endeavour drawn from the journals kept by the several commanders and from the papers of Joseph Banks ...*, 3 volumes, London, 1773.

Price, A.G., *The Explorations of Captain James Cook in the Pacific As Told by Selections from His Own Journals, 1768–1779*, Dover Publications, New York, 1971. (An illustrated, accessible, paperback edition.)

James Cook also wrote some sets of *Sailing Directions* on the basis of his early surveying work in Canada, and a short note on his observations on the eclipse:

Cook, J., 'An observation of an eclipse of the sun at the Island of Newfoundland, August 5, 1766, with the longitude of the place of observation deduced from it, communicated by J. Bevis, M.B., F.R.S.', *Philosophical Transactions of the Royal Society*, Vol. 57 (1767), 215–216.

Chronology

1728	Born 27 October at Marton-in-Cleveland, Yorkshire
1736	Moved to Great Ayton; attended Postgate School
1745	Apprenticed to Mr William Sanderson; after eighteen months went to Whitby; apprenticed to John Walker, shipowner
1752	Became mate of *Friendship*
1755	Volunteered for Royal Navy as able seaman; soon became master's mate on the *Eagle*
1757	Qualified as master, joined *Solebay*, later *Pembroke*
1758	Sailed for Canada under Admiral Boscawen. Survey of St Lawrence River, Canada
1759	Transferred to *Northumberland*; survey of coasts of Nova Scotia and Newfoundland; commended for his work
1762	Married Elizabeth Batts, at Barking, Essex, 21 December
1763	Appointed Surveyor of Newfoundland. Continued survey of Newfoundland coast, as commander of *Grenville*
1766	August, made observations on eclipse of the sun; came to the notice of the Royal Society
1768	Appointed to command expedition to observe the transit of Venus from the South Seas. Commission as lieutenant signed; on his advice a Whitby collier chosen and renamed *Endeavour* prior to voyage to Pacific to ...
1769	... observe transit of Venus in Tahiti, June; explorations of New Zealand followed
1770	20 April, first sighting of Australia, Point Hicks, Victoria; explorations of east coast of Australia; claimed for British Crown at Possession Island, Queensland, 21 August
1771	Returned to England, via Batavia
1772–5	Second Pacific voyage aboard *Resolution* with *Adventure*; crossed Antarctic Circle three times. Promoted post-captain on return
1776	Received Copley Medal and Fellowship of the Royal Society
1776–9	Third voyage, in search of the North-West Passage in *Resolution* with *Discovery*. Halted by ice at 70° 44′N
1778	Visited Hawaii (Sandwich) Islands
1779	Killed in Hawaii, 14 February. Remains buried at sea, 22 February

This chapter has been written from two of the principal loci of Captain James Cook's life and work: North Yorkshire and Australia. Jill Rutherford lives near Whitby, close to where James Cook worked as a young man; she is an educational consultant and works for the International Baccalaureate Organisation and the University of Hull. Patrick Armstrong teaches geography at the University of Western Australia.

Owen Lattimore

1900–1989

Gary S. Dunbar

Courtesy of the Ferdinand Hamburger, Jr Archives of The Johns Hopkins University, Baltimore, USA

The distinguished American Sinologist and Mongolist Owen Lattimore made noteworthy contributions to the geography and culture history of Inner Asia in his long career as a journalist and university professor, but, ironically, his most enduring fame derived from his persecution at the hands of right-wing politicians in the early part of the Cold War. Just as his writings continue to inspire and inform scholars who seek an understanding of Asian affairs, the moral lessons of his life story have not lost any of their significance. This essay will emphasize Owen Lattimore's geographical interests and activities.

1. Education, Life and Work

Owen Lattimore was born 29 July 1900 in Washington, DC, the son of David and Margaret (Barnes) Lattimore. David Lattimore (1873–1964) attended normal school for a year and then taught Latin at Washington High School from 1893 until 1901, when the family moved to China, where David was employed for the next two decades as a Professor of English in three institutions (Nanyang College, Shanghai, 1901–5; Chihli Provincial College, Paotingfu, 1905–13; and Peiyang University, Tientsin, 1913–21). In February 1922 he assumed a professorship in Far Eastern Civilizations (teaching the Chinese language) at Dartmouth College, Hanover, New Hampshire, and retired as Professor of History in 1943. David's older sister Mary, a friend and disciple of Alexander Graham Bell, had gone to China in the 1890s under Presbyterian auspices to teach deaf and dumb children, and she persuaded her brother to apply for a job in the Chinese educational service that was expanding rapidly after the end of the Boxer Rebellion.

Owen and his four siblings were educated at home and were widely read in classical literature and in European culture generally. Owen had a particular interest in works of geographical discovery and exploration, and he was attracted to the writings of the French-American explorer Paul Du Chaillu (1831–1903). Like his sisters, Owen had a great love of poetry, but in his knowledge of Greek and Latin literature he was to be surpassed by his younger brother Richmond (1906–1984), who went on to become a renowned classicist at Bryn Mawr College in Pennsylvania. Owen's youth was punctuated by some remarkable events, such as the revolution of 1911, when he witnessed impaled heads on telephone poles in Peking. David Lattimore had made the acquaintance of W.W. Rockhill (1854–1914), the American Minister to China from 1905 to 1909, who had been an explorer in Central Asia (and who became a charter member of the Association of American Geographers in 1904).

In the spring of 1912, David Lattimore, who was preparing to leave Paotingfu for a university position in Tientsin the next year, decided that he did not want his children to grow up in China, and so he packed his wife and their five children, ranging from six to thirteen years of age, off to his beloved Europe (which he himself had not yet visited). As Owen later said, 'He [Father] wanted us to spend two or three years in Switzerland, to get a good start in French and German, then go on to America in time to adjust and prepare for college. After college we could each make our own decision: stay in America, go back to Europe, or return to China. As it turned out, my elder sister and I, born in America, were the only ones to go back to China; the other three, born in China, never saw China again.'

The family, *sans père*, sailed to Europe, where they spent the next two years in Lausanne, Switzerland. Owen, whose only memories had been of northern Chinese landscapes, did not find anything very remarkable about the sea voyage between China and Italy, but he was startled to see forests growing on flattish land when the family travelled by train from Genoa to Lausanne. The landscape was an utter contrast with China, where trees were sacrificed in order to extend the arable land for the dense population. This was his first vivid lesson in comparative geography.

In Lausanne, Owen attended the Collège Classique Cantonal for two years, working very hard to get his French and German up to the level where he could compete with his Swiss classmates. Also in Lausanne at that time was David Lattimore's younger brother Alec, 'the Bohemian of the family', who had been travelling in Europe for several years, learning 'to speak excellent German and appalling but fluent French'. In Paris Alec had met a wealthy Polish family, who hired him as a tutor for their sons. He was in Lausanne to keep watch over one of the sons, who was enrolled in an expensive boarding school. When the war clouds were gathering in the summer of 1914, Alec took Owen with him to Oxford, where his Polish charge was to enrol in the university. Mrs Lattimore soon followed with the other children and then departed for China, leaving Owen in the care of Uncle Alec. It was decided that Owen could not receive proper schooling in China, and as there were no close relatives in America who could look after him, he was stranded in England for the duration of the war. David found Alec a teaching job in China, and the latter arranged for his nephew to live with an English friend in Cumberland and to attend the nearby St Bees School. (In 1995 Owen's son David described St Bees as 'probably the smallest, most remote, and cheapest of ancient and honorable public schools'.) After a miserable experience as a 'day boy' travelling several miles to school, sometimes by train, sometimes by bicycle, Owen became a boarder at St Bees and began to spend his holidays in Scotland with one of his father's Paotingfu colleagues, a chemist named James Henderson. The Hendersons lived in West Kilbride, near Glasgow, and later moved to the Manchester area. Lattimore described Henderson as a genuine example of the proverbial 'dour Scot', quite deaf and not given to small talk, 'but in the Scottish tradition he was a humanist as well as a scientist'. 'There was good literature on his bookshelves, and it was also there that I found the books that first interested me in geography (Elisée Reclus) and anthropology (E.B. Tylor).' He was also introduced to the works of Ernest Renan and then to the writings of his father's hero, T.H. Huxley.

Owen Lattimore spent almost five years at St Bees School, leaving on his nineteenth birthday. The school left an indelible mark on him, not least in stamping his speech and writing with touches of British usage that would set him apart from other Americans. He excelled in literary studies and was the principal contributor to a new school paper that he had helped to found (*The Mitre*), but he was not able to win a scholarship to Oxford, and so he returned to China in the fall of 1919 to begin a career in business and journalism.

> What desolated me was that I had no clear idea, no real interest in life except getting to Oxford. I had to survive about six years of boredom and discontent, in which I refused to admit that I was really having a pretty good time, before I was ready to admit that the Oxford of the 1920's would very likely have been the ruin of me, and a lot longer than that before I understood that my six wasted years had been worth as much as ... a multiple First Class degree in politics, economics, history and sociology. At Oxford I probably would have foundered because of immaturity. I would have been attracted, in poetry, to the worst that was being written there in those years; in prose, to the flashy and meretricious. In politics and economics I could have come out either on the sentimental left or the romantic right; in either case, my feet would have been planted firmly in the air. ('Happiness Is Among Strangers', p. 71)

After the long separation, his family received him warmly in their new home in Tientsin. His father reviewed his work at Lausanne and St Bees and concluded that Owen had achieved a standard equal to that of an American student who had finished the second year at a good college or university, superior in literary pursuits but deficient in the sciences. Through an American friend, Owen secured a position in the Tientsin office of the British export firm of Arnhold and Company. His first assignment was a boring job in the Piece Goods Department, dealing with cotton textiles, but after a few months he was sent to Shanghai, where he spent a year in the firm's insurance office. On his return to Tientsin he took a job as subeditor of a newspaper, *The Peking and Tientsin Times*. David Lattimore had been offered a professorship at Dartmouth College, but he did not assume the post until February 1922. He sent his wife and children back to the United States a year early to establish them in American schools. Owen spent that year living with his father and Uncle Alec in their quarters in Peiyang University, thus cramping his social life but bringing him closer to his father ('though much more intellectually than personally'). 'He was the most learned man I have ever known', said Owen, but 'on all things personal he guarded his reserve closely'. Politics did not much interest Owen in those years, but he managed to fill in his spare time with reading and athletic activities (rowing, rugby, riding, and polo). He began reading Swinburne and G.D.H. Cole but lost his former interest in Thomas Hardy, G.K. Chesterton and Hilaire Belloc. His avid interest in learning Chinese and in travelling in the interior set him apart from the majority of Europeans, who were content to remain in the coastal cities in sublime ignorance of the local culture.

Back at work for Arnhold, both in Tientsin and in Peking, Owen found himself dealing in export items such as fibres (wool, camel hair and cashmere) and carpets. His close contact with Chinese traders and caravan men gave him remarkable insights into Chinese culture ('a marvelous way of marinating myself in the language'), in contrast to older Europeans who had spent several decades in China but were still unable to speak enough Chinese to order a meal.

According to custom, a foreigner sallied into the interior ... with lots of 'face' or self-importance: an English-speaking interpreter, a pidgin-English-speaking servant to cook for him and wait on him, a camp bed, changes of clothing, cases of tinned food and bottles of liquor. On arrival, an inner and outer room would be cleared for him at the best inn, local gapers would be shooed away, and he would hold court, receiving the agents of the Tientsin compradors and emissaries of the local officials as if, besides being the representative of a single foreign firm he were a kind of civil servant on tour. When I turned up with no interpreter, no servant, no provisions, our local agents were flustered, but I insisted that I would just bunk in with them and eat their food ... It was clear at once that I had done the right thing. It is when you are sitting around gossiping after hours, talking a mixture of shop and local gossip and general news that you learn what is going on and how things work. ('Happiness Is Among Strangers', pp. 110–11)

In 1924 Lattimore was transferred to Peking, and there the next year he met his future wife, Eleanor Holgate. Eleanor was the daughter of Thomas Holgate, a Professor of Mathematics at Northwestern University (in Evanston, Illinois), who had been invited in 1922 to spend a year as a Visiting Professor in the University of Nanking. Becoming enamoured of China, and especially of Peking, Eleanor seized the opportunity to return in 1925 as Secretary in the Peking Institute of Fine Arts. Owen immediately took to Eleanor, who was six years his senior; they were married on 4 March 1926.

There followed an amazing wedding trip, which Lattimore's biographer Robert Newman has called 'a honeymoon for the ages'. First, Owen was to go westward through Inner Mongolia to Sinkiang by camel caravan, and Eleanor would wait and take a faster and safer trip northward from Peking and then westward on the Trans-Siberian Railway, meeting up with her husband at Semipalatinsk, some 400 miles from the border of Sinkiang. Owen planned to depart in March, soon after the wedding, but his camels were commandeered by a Chinese warlord, and he was not able to start his journey until August. He arrived in Urumchi in January 1927 and sent telegrams to Eleanor, who began her train journey in February. Owen was not given permission to cross the border to meet her in Semipalatinsk, but he reasoned correctly that she would somehow evaluate the situation and find him in the Chinese border town where he was waiting for her. After a difficult seventeen-day sledge passage, she joined her husband in late March, and they began their six-month wedding trip through Central Asia. In September they crossed high passes in the Karakorum Range and arrived in Kashmir. From India they went to Italy, where they spent the winter in Rome writing accounts of their travels, using the library resources of the Italian Geographical Society. When Owen completed his manuscript they left Italy, going first to Paris and then to London. In Paris Owen met with a chilly reception from the famous Sinologist Paul Pelliot, in contrast to the warmth with which he was received in London by the naturalist Douglas Carruthers, whose two-volume work *Unknown Mongolia* (1913) was so much admired by Lattimore that he carried it with him through Central Asia. Owen quickly found a publisher, and his book *The Desert Road to Turkestan* was issued in 1928. Lattimore was encouraged to join the Royal Geographical Society, to deliver a lecture there, and to publish his first article in the *Geographical Journal* ('Caravan routes of Inner Asia'). Some of his professional traits were established at that time. Following Carruthers' example, Lattimore maintained a lifelong openness towards young scholars and their work, sometimes even to the extent of writing patient replies to people who might kindly be described as living perilously close to the edge of reality. Lattimore later said that his lecturing style derived from advice given him by the Secretary of the Royal Geographical Society, A.R. Hinks, just before that first lecture. Lattimore described the situation to Alfred Steers in a letter in 1981:

> When I turned up for my first ever RGS lecture, he [Hinks] asked me if I had a text. I said I had. He held out a hand. 'Give it to me,' he said. 'I shall now impound this and you will speak without a text. Lectures should be spoken, not read. People understand the spoken word and the written word quite differently.' I obeyed orders and have since never lectured from notes or a text, writing a text only if there is to be later publication. (Library of Congress, Manuscript Division, Lattimore Papers, Box 19, Folder 9)

Owen finally returned to the USA in 1928. Isaiah Bowman, the Director of the American Geographical Society, had a special interest in frontier regions ('the pioneer fringe'), and he was taken with Lattimore's proposals for the study of Inner Asia. Bowman used his considerable influence with the Social Science Research Council to help Lattimore get a grant for a year of post-graduate study (without having done any undergraduate work) in anthropology at Harvard University (1928–9) and a second year of travel and study in China. The SSRC stipend and successive grants from the Harvard-Yenching Institute and the Guggenheim Foundation enabled the Lattimores to return to China for another four years of study and travel, during which Owen began to learn the Mongol language and conducted most of the travel that would form the basis for *Inner Asian Frontiers of China* (1940).

In 1928 Owen Lattimore sold his first article (of four) to the *National Geographic Magazine* for the very considerable sum of $800. Initially, the article bore the title 'Innermost Asia', but when it was learned that Aurel Stein was about to publish a book with the same name, it was changed to 'The Desert Road to Turkestan' and appeared in the June 1929 issue of the *Magazine*. Over the years Lattimore enjoyed cordial relations with the National Geographic Society. He perfectly well understood their objectives and tailored his writing accordingly. He saved his political opinions for more appropriate publication outlets.

When the Lattimores arrived in Boston in 1928, they made the acquaintance of a remarkable couple, Robert and Katharine Barrett. Robert Barrett (1871–1969), the scion of a well-to-do Chicago family, had been a student of William Morris Davis at Harvard. He had been a charter member of the Association of American Geographers in 1904, and by the time he died 65 years later, Barrett had outlived all the others. As a man of means, Robert Barrett never had to seek employment, and so he spent his life travelling and following his own interests. He is perhaps best known to geographers for his self-financed expedition with Ellsworth Huntington to Central Asia in 1904–5. Barrett took an immediate liking to the Lattimores and gave them considerable financial support over the years. The Barretts were also to lend great moral support to the Lattimores, especially during the time of the McCarthy witch-hunt in the early 1950s.

From a base in Peking, Owen Lattimore was able to travel widely through Inner Mongolia, Sinkiang and Manchuria in the years 1930–3. By the end of his second Guggenheim year, in 1933, he was ready to return to the United States and seek suitable employment. On the recommendation of H.G.W. Woodhead, the editor of *The Peking and Tientsin Times* under whom he had worked a decade earlier, Owen was made the editor of *Pacific Affairs*, the journal of the Institute of Pacific Relations (IPR), which was based in New York. After spending several months in

New York learning his new job, Owen returned to Peking in 1934 and spent the next four years editing the journal from there. Among his American friends in Peking during that period were Joseph Stilwell, John King Fairbank, Edgar Snow, John Stewart Service, H.G. Creel and Carl Bishop. Bishop was a field archaeologist whose empiricism had a great effect on Lattimore's thinking. He was to have a profound influence in shaping Lattimore's *Inner Asian Frontiers of China*.

Lattimore's acceptance of articles for *Pacific Affairs* on the sensitive topics of colonialism and imperialism and the IPR's attempts to bring Russians into their programme had the delayed effect of causing political attacks on Owen Lattimore almost two decades later, in spite of the fact that the Russians were his biggest critics and had boycotted, or tried to sabotage, the Institute's conferences and other activities. In 1936 Edward Carter, the Secretary-General of the IPR, instructed Lattimore to spend some time in Moscow en route to the annual Institute conference that was to be held in Yosemite Park (California) in the summer. After the conference Lattimore spent twelve weeks in London studying Russian, but he was denied a visa to return to the USSR and so returned to China in the spring of 1937. In July of that year the Japanese took over Peking, and Lattimore soon saw that it would be difficult to continue editing *Pacific Affairs* from there. The Lattimores embarked for the United States in December 1937 and spent the first half of 1938 in California, where Owen continued editing the journal and working on the manuscript of *Inner Asian Frontiers of China*.

Before leaving China, Lattimore had written to Isaiah Bowman, who had assumed the presidency of the Johns Hopkins University in 1935, about academic prospects for a China specialist. Earlier, when Bowman had been the Director of the American Geographical Society, the Society had commissioned Lattimore to write a book on China's frontiers. The work had already taken so long that Lattimore now felt that he should return the $500 advance. He wanted to rewrite the book completely, culling the 'dross' or 'nonsense' from it, and shortening it to one volume from three.

Bowman replied to Lattimore: 'There is only one university for you to consider with respect to the future and that is Johns Hopkins. I have had you on my mind ever since I came here.' In January 1938 Bowman sent the following telegram: 'I can now offer you a lectureship in the Page School on half time at three thousand dollars annually to begin whenever you choose to come.' Lattimore accepted and began his quarter-century association with Johns Hopkins in September 1938. His half-time appointment in the Walter Hines Page School of International Relations allowed him to continue editing *Pacific Affairs*. The editorship lasted until 1941, but he assumed a full-time appointment as Director of the Page School in 1939. His book *Inner Asian Frontiers of China* was published in 1940 by the American Geographical Society, received highly favourable reviews, and to this day – six decades later – remains his most durable work.

During his first year at Hopkins Lattimore attended meetings of the American Philosophical Society in Philadelphia, and at one of the meetings he met the Arctic explorer Vilhjalmur Stefansson (1879–1962), who was to become a very close friend. The two travellers were both great raconteurs, and they enjoyed each other's company immensely, even though their experiences had been gained in very different milieux.

In the early 1940s Lattimore took leave from Johns Hopkins to serve the United States and China in various capacities. With the support of Isaiah Bowman and the White House aide Lauchlin Currie, he was appointed personal advisor to Chiang Kai-shek in 1941. He was not paid by the US government, as many people later assumed, but by the Chinese Nationalist government. He accepted a six-month assignment in June 1941 (later extended) and arrived in Chungking, Chiang's wartime capital, a month later. Lattimore's major duties seemed to consist of 'drafting and revising Chiang's many appeals to Roosevelt' for American aid in the Chinese war with Japan. Lattimore had long been an admirer of Chiang, but he became critical of the way the Generalissimo handled the internal struggle with the Communists. His US leave was delayed until mid-January by the sudden American entry into the Pacific war, and he arrived in the US on 8 February 1942, not returning to China until the end of September. He then tendered his resignation to Chiang because he wanted to take up a new assignment as Director of the Pacific Bureau of the US Office of War Information in San Francisco. He left Chungking on 19 November, taking up his new duties in San Francisco in late December 1942. He continued in that position for fifteen months.

Not long after he returned home to Baltimore in March 1944, he was persuaded to join US Vice-President Henry Wallace on a two-month tour of Siberia, Soviet Central Asia, China and the Mongolian People's Republic. The mission departed from the United States on 20 May 1944, travelling first to Siberia, where they were well received by the Russians, who managed to deceive the Americans by showing them some slave labour camps disguised as 'Potemkin villages'. Wallace's and Lattimore's subsequent favourable descriptions of Soviet life gave American anticommunists more ammunition to use against them a few years later. After touring Siberia and Soviet Central Asia, the Wallace mission spent two weeks in China and two days in the Mongolian People's Republic. This was Lattimore's first visit to the MPR (Outer Mongolia). Although he was fluent in Mongolian, all of his experiences with Mongol culture had been acquired in the Chinese provinces of Inner Mongolia, and he had previously not been able to acquire the necessary Russian permission to visit their satellite republic.

Lattimore returned to Johns Hopkins but not to a cloistered academic life. In the postwar era there was a greater demand than ever for his writing and lecturing skills outside the university, particularly when the Chinese resumed their civil war. His lack of formal academic qualifications also meant a lack of allegiance to a single discipline. He was a geographer, to be sure, but he was even closer to the disciplines of history and political science. Countering those detractors who said that he was only a journalist and therefore not a scholar, one can give ample evidence that he was both scholar *and* journalist. More than most academics, he wrote for the popular press and gave numerous lectures outside universities, partly to gain a greater forum and partly to augment his salary. His title, Director of the Walter Hines Page School of International Relations, obscured the fact that the School operated on a shoestring. For example, in the academic year 1947–8, the budget of the Page School was only $9856, including $6000 for Lattimore as Director and $2080 for a secretary. Lattimore supervised the programme, but the other participating professors came from other departments. The 'courses and seminars bearing on international relations' (1948–9) included Lattimore's courses in history and political science, courses in geography by Ernest Penrose and Robert Pendleton, and a political geography seminar taught by Lattimore and Penrose. Disciplinary boundaries have perhaps been rather vague and permeable in the Johns Hopkins University, making it just the sort of place where an independent thinker like Lattimore could flourish, at least in the period before some politicians began to try to extend their power into the halls of academe.

The painful and protracted persecution of Owen Lattimore by anti-communists in the United States Senate in the early 1950s has been fully documented in many books and articles, most notably by Robert Newman in his 1992 biography of Lattimore, and so will be briefly treated here. After the Chinese Communists won final victory over the Nationalist forces in 1949, there were many influential Americans, mostly conservative politicians and newspaper publishers, who felt that the United States must have been responsible for the betrayal of its wartime ally Chiang Kai-shek and the subsequent 'loss' of China. The cudgels were enthusiastically taken up by political opportunists such as Richard Nixon and Joseph McCarthy who saw anti-communism as a 'ticket to success'. Although McCarthy and the other 'moral entrepreneurs' (to use Lionel Lewis's apt phrase) really wanted to bring down the most highly placed members of the Democratic administration, such as Harry Truman, Dean Acheson and George Marshall, they had to be content with trying to destroy the careers of some administrators and scholars who were less well placed. They picked on Owen Lattimore because of his expertise on China, his closeness to Chiang, and his forthright commentary on Soviet and Far Eastern politics, including the growing hostility between the two Koreas that would lead to outright war in June 1950. Since the United States had failed to influence events in China and Korea, McCarthy and the other demagogues drew the (to them) logical conclusion that American Communists and 'fellow travellers' must be to blame. They were incensed by the fair treatment given to Lattimore by Senator Millard Tydings in the initial Senate hearings in the spring of 1950 but were emboldened by Tydings' defeat and the election of Richard Nixon to the Senate in the autumn. McCarthy even paid Lattimore a left-handed compliment when he called him 'the top Soviet agent in the United States'. While some China experts who were criticized by Senators McCarthy and Patrick McCarran swallowed their pride and tried to avoid antagonizing the inquisitors, Owen Lattimore gave back as good as he got. Among the witnesses who testified against him were former Communists (and former friends) such as Karl Wittfogel and Freda Utley, who fed the McCarthyites what they wanted to hear in order to improve their own positions. Wittfogel had spent nine months in a Nazi concentration camp and was determined to stay out of any such camp that might be established in postwar America. Turning on Lattimore would help to demonstrate his *bona fides*. Utley claimed that her experience in Communism had given her the ability 'to detect the cloven hoof' and that Lattimore's guilt was therefore quite obvious to her. George Carter, a Hopkins geographer, was also critical of his former friend. In the end, Lattimore's wit and sarcasm, combined with his superior knowledge of China, so inflamed his tormentors that they induced the Justice Department to indict him for perjury in 1952 and 1954.

Another geographer whom McCarthy attempted to smear was Peveril Meigs (1903–1979). The Senator lost interest in Meigs when he thought, mistakenly, that the Department of the Army had fired him for appearing on McCarthy's hit list; in fact, Meigs continued to head the Earth Sciences Research Division of the Army Quartermaster Corps until his retirement in 1965.

Under pressure from the Johns Hopkins trustees, Lattimore went on leave from the university at the end of 1952, and he did not resume his academic post until the second perjury indictment was dismissed by Judge Luther Youngdahl in 1955. McCarthy's behaviour not only in the Lattimore case but also in the subsequent Army trials in 1954 earned him the opprobrium of many Americans, liberals and conservatives alike. Thus it might be said that Lattimore had performed a service to his nation, even though it had cost him dearly. He is credited with coining the term 'McCarthyism' in the pejorative sense in which it has been used ever since, despite McCarthy's vain attempt to turn it around and equate it with patriotism. Unfortunately, McCarthyism did not die with McCarthy in 1957. It keeps returning in new guises, but it is hoped that the dark times of the early 1950s will not be revisited. The moral lessons of Owen Lattimore's experience must not be forgotten.

The experiences of Owen Lattimore and other academics in the parlous period of Red-baiting and loyalty oaths have been important in strengthening the cause of academic freedom in the United States. Many university presidents and boards of trustees had simply capitulated to right-wing pressures. Some of them sheepishly rebounded and began to assume a firmer stance in the affirmation of academic principles. By and large, the Johns Hopkins administration behaved quite well throughout this period, although Lattimore might have helped his own cause if he had practised a little more diplomacy in his relations with university colleagues and officials. University boards of trustees usually consist of business people who may be university graduates themselves but rarely share the academic staff's outlook and values. There is a tension between the two groups, with the university presidents being caught in the middle.

When the second perjury indictment was rejected in 1955, Owen Lattimore returned to Johns Hopkins but was not given a hero's reception. The governing board of the University closed the Page School, partly to indicate its coolness towards his return, but he was invited to join the Department of History by its chairman, Sidney Painter, a political conservative who had always stood for academic freedom. Although Lattimore was thereafter given considerable latitude in crossing disciplinary boundaries in the characteristic Hopkins fashion, he was not so happy with his diminished role in his final eight years in the University. His relations with the Geography Department improved considerably after his friend Gordon Wolman returned to Johns Hopkins in 1958 as chairman of the department.

Despite his long association with The Johns Hopkins University, Owen Lattimore was happy to be called to the University of Leeds in 1963 to head the newly established Department of Chinese Studies. He had been urged to accept the Leeds offer by his friend Laurence Kirwan, Director of the Royal Geographical Society. 'Of course, you must take the job at Leeds', wrote Kirwan in 1962, 'good for you, good for us, good for the country, good for geography, for we shall then have one professor in that field worth the name (and salary).' However, Lattimore's small department was mainly concerned to promote the study of the Chinese and Mongol languages and culture.

When Lattimore retired from Leeds in 1970, the Lattimores planned to return to the United States to a house that was being built for them in Virginia. What should have been a happy homecoming turned to grief when Eleanor died in New York after being stricken with a pulmonary embolism just as the plane was landing at Kennedy Airport. Owen was greatly dispirited, because they had been so close in their 44 years of marriage. Eleanor Lattimore (1894–1970) was a partner in much of her husband's work. After publishing her own remarkable book *Turkestan Reunion* (1934), she subordinated most of her activities to those of her husband. She edited much of his writing, and they collaborated in producing two popular books, *The Making of Modern China* (1944) and *Silks, Spices and Empire* (1968). These books, her extensive travels, and her membership in the Society of Woman Geographers all qualify her to be considered a geographer in her own right. The Lattimores had one child, David, who was born in Peking in 1931 and now, as a Professor Emeritus in Brown University, has the distinction of being the third generation of Lattimore men to serve as a professor in a major American university.

In the first dozen years or so after Eleanor's death, Lattimore was almost constantly on the move. Without her, he could not feel at home in the Virginia house, which was his base from November 1970 until February 1973, when he moved into a rented apartment in Levallois Perret, a suburb of Paris. From Paris he would venture forth several times a year, most often to England, and he bought a house in Leeds in 1974 because he wanted to be of further assistance to his old department. He would attend conferences all over the world, lecture on the American university circuit, and visit Mongolia almost every year. He sold his Leeds house in 1979 and moved to an apartment in Cambridge. He described his Cambridge years as 'blissful', because he enjoyed being in closer touch with scholars with whom he was especially compatible, such as Joseph Needham, Edmund Leach, Joan Robinson, Moses Finley and E.H. Carr. He did not seem to be close to the Cambridge geographers, however, with the exception of B.H. Farmer. He frequently visited London, attending meetings of the Royal Geographical Society and the Royal Central Asian Society.

On his 77th birthday Lattimore began to write his memoirs, to which he gave the title 'Happiness Is Among Strangers'. He had lived about two-thirds of his life outside the United States, and he was comfortable among people in whatever part of the world he found himself. His test of happiness was simply this: 'If you are living in a country not your own, you have to gain access; you have to work at it; and to work at a problem, get the feel of it and succeed – that is happiness.' The writing went slowly, and he had produced only 140 typewritten pages (covering the first 25 years of his life) by 1980, when he submitted them to an editor at the Oxford University Press. The Press expressed interest but wanted to see a more complete version before they could make a serious commitment. He soon despaired of completing the work and was urged by his long-time assistant Fujiko Isono to concentrate instead on writing a book about his interaction with Chiang Kai-shek in World War II. He agreed, but after writing only a few pages he gave up and allowed Isono to tape lengthy interviews instead. Her draft of the manuscript was ready by 1984, and it was given to John DeFrancis for editing. The book was published posthumously in 1990 by the University of Tokyo Press. Its title, *China Memoirs: Chiang Kai-shek and the War Against Japan*, disguises the fact that it contains considerable material on Lattimore's life before and after his relations with Chiang.

By 1985 Lattimore's family and friends were worried about his ability to manage alone; so his son persuaded him to leave Cambridge and join him in Pawtucket, Rhode Island. His health was then reasonably good for a man of his age, and he managed to keep up his interests in world affairs through reading and conversation. His biographer Robert Newman was able to visit him more often.

In 1986 the Association of American Geographers, 'very conservative and long uneasy about Lattimore' (Newman's words), gave him an Honors award *in absentia*. Apart from a tiny minority, how many American geographers were 'uneasy about Lattimore'? It is my observation that he was always appreciated by all those geographers who knew him personally or through his writings. Scholarly societies are by definition 'conservative' in the best sense, in that they are careful to conserve the best from the past while expanding into the future, avoiding faddishness and the radical extremes on the political right as well as the left. Owen Lattimore was never a member of the Association of American Geographers. He had been a member of the Royal Geographical Society from 1928 onward, except for the period 1940–9.

Just before his 87th birthday Owen Lattimore suffered a stroke that severely impaired his ability to speak, and he was forced to cancel a projected trip to Mongolia. He continued to read widely, but his range of activities was narrowly circumscribed. He died of pneumonia in the early morning of 31 May 1989 in Miriam Hospital in Providence, Rhode Island.

2. Scientific Ideas and Geographical Thought

Owen Lattimore possessed no earned degrees, but his non-journalistic works met or exceeded the standards of academic scholarship. One can well imagine, for example, that his 1933 monograph on the Gold Tribe would have been perfectly acceptable as an MA thesis in anthropology, or that *Inner Asian Frontiers of China* (1940) could have earned him a doctorate in geography or history. Fortunately, in that era he was not required to have paper credentials, and so he was not tied to any particular field of specialization. Unlike most academics, he was not beholden to any mentor or school of thought but was free to roam widely through the literature of the humanities and social sciences. He could easily absorb new ideas and jettison older ones that no longer appealed to him. He could voice approval of certain ideas of Karl Marx, Thomas Huxley or Arnold Toynbee, for example, but just as easily oppose some of their other notions.

In his early travels and writings, Owen Lattimore paid much attention to the ideas of the American geographer Ellsworth Huntington, especially as expressed in his *The Pulse of Asia* (1907), but Huntington's determinism was rejected when Lattimore reasoned that the causes of desiccation in Central Asia were more cultural than physical. After a phase in which he took inspiration from Oswald Spengler's *The Decline of the West* (1918–22; English translation, 1926), he passed on to the Marxist ideas of Karl Wittfogel about 'Oriental despotism' – ideas that were essentially the same as those of the non-Marxist archaeologist Carl Whiting Bishop. Lattimore was never a Marxist nor a Communist. As he explained his philosophical position:

> By the time that events began to bring me into contact with Marxists of various sects I was already in my middle thirties – past the age of youthful apocalyptic conversion to any doctrine and disposed instead, like [Thomas] Huxley, 'not to care much about A's or B's opinions, but rather to seek to know what answers he had to give to the questions that I had to put to him'. (*Studies in Frontier History*, p. 28)

Much of Lattimore's published work had to do with frontiers, particularly the northern and western (but not the southern) borderlands of Han China. This interest continued throughout his entire life, marked by the monuments of *Inner Asian Frontiers of China* (1940) and *Studies in Frontier History* (1962), the latter a collection of papers originally published between 1928 and 1958. Those interested in the intellectual evolution of his frontier concepts should not confine their study to these books, however, but should consult also his later papers and his autobiographical ruminations of the 1970s and 1980s. Lattimore's ideas are ever-new, and scholars are afforded endless opportunities for reinterpretation of his place in Asian historiography.

To geographers and other scholars, Lattimore's most substantial and enduring work is *Inner Asian Frontiers of China* (1940), a work encouraged by Isaiah Bowman and published by the American Geographical Society. In his introduction to the latest reprint, Alastair Lamb has praised Lattimore and remarked particularly on his methodology, which was long on personal experience and short on archival search. In Lamb's view, Lattimore 'showed how sophisticated models could be constructed which were both useful and intellectually satisfying. The book is more a demonstration of method than a work of reference ...' 'The methods exploited by Lattimore' were 'first ... a thorough knowledge of the ground derived from actual experience' and, second, 'a deep knowledge of at least some specific periods of history'. Lamb continues:

> The devising of historical models can produce writing of a turgid obscurity or jargon-ridden tedium, the one usually associated with Teutonic scholarship at its Spenglerian worst and the other with the more obsessive Marxists. Lattimore writes with a style which leans towards neither of these extremes. (*Inner Asian Frontiers of China*, 1988, p. viii)

For close examination of Lattimore's place in frontier studies, I turned to Julian Bishko, a Turnerian historian of the medieval Iberian frontier. Professor Bishko, who has read widely in the literature of the American, African, Asian and Australian frontiers, has shown that Lattimore used 'frontier' more in the European sense of 'border' or 'borderland' than in the way in which Frederick Jackson Turner and subsequent American historians have commonly employed it. As Professor Bishko wrote in a letter to me (2 September 1996):

> I have always had difficulties with Lattimore's use of the vocable 'frontier' because it constantly passes from the American Turnerian sense over to what Turner carefully distinguished therefrom: the European meaning of 'border' ... Lattimore seems to me essentially, and in most of his writings, a political scientist, interested in power politics, in shifts of territorial domination, and, specifically, in the fluctuating fortunes of regions caught between the power thrusts of surrounding Great Powers...
>
> He is primarily focused on such topics as emigration, settlement, stages of social, economic and institutional development; he is not dealing with the once-and-for-all movement of a more developed society into a wilderness or thinly settled or uninhabited territory, which, once subject to the 'frontier' process, becomes an integral part of the metropolis and the base for the next roughly comparable advance of the 'frontier' ...
>
> There is ... the contrast between the Turnerian and American notion of a frontier area penetrated and settled once and for all ... [and] Lattimore's far larger framework – a *longue durée* in the style of the *Annales* school but one even more extended than they envisage – one with cycles and stages of organic change in the tradition of Huntington, Spengler et al. ...

In using Turner as a standard in frontier studies, Bishko is by no means trying to denigrate Owen Lattimore, whom he praised for his 'brilliant insights and interpretations'. Again we see that the American-born Lattimore was essentially a European in his intellectual formation. Much of his character was shaped by his English school experience, and he probably would have been better suited to the life of an Oxbridge don than to that of an American university professor in uncomfortable proximity to Washington.

Owen Lattimore was an unabashed romantic. From first to last, he preferred the steppe to the sown, the herdsman to the farmer, the nomadic encampment to the city. Although he came late to the study of the Mongol language and later still to travel in the Mongol heartland (Outer Mongolia), he found his real home north of the Great Wall. An adept horseman in his young adult years, he retained a keen interest in Mongol horses and horsemanship to the end of his life. He developed deep attachments to the horses and camels that facilitated his travels and even chided himself for failing to describe their individual characteristics in his books. As he said to J.O. La Gorce of the National Geographic Society (letter, 27 September 1940),

'It is depressing to think that I should have so ungratefully forgotten the personalities of those camels with which it was such fun to travel'. Lattimore's 'ethnographic eye' – akin to the 'morphologic eye' that geographers are supposed to cultivate – gave him keen insights into the lives of the nomads. He continually refined and rearranged his views on the nature of Inner Asian nomadism. In one of his last publications ('Herdsmen, farmers, urban culture') he admitted to a change in his earlier view that had been derived from his experiences of approaching the Mongol culture area from the south: 'In recent years ... I have come to the opinion that the world of nomadic pastoral Asia can be seen in better proportion and perspective if looked at from the Middle East and Central Asia than if studied from the Great Wall of China.' If only he had had time left to expand on these views and incorporate them into a new edition of *Inner Asian Frontiers of China*!

Another late paper that should still have appeal to geographers is 'The periphery as locus of innovation' (1980). Here he challenges the earlier, more traditional view that innovations usually originate at or near the centre of a culture and 'radiate outward to its frontiers, where they lag behind developments at the centre in retarded forms'. Lattimore suggests an alternative view, based on examples from Europe and North America but especially from China, of the vitality of reverse movements from periphery to centre. Geographers do not have to look very hard to find some real nuggets in Lattimore's ruminations and speculations.

3. Influence and Spread of Ideas

As a quasi-geographer not fully committed to any academic discipline or even to the academic life generally, Owen Lattimore was in a position to influence people in positions of real power in society. His numerous speaking engagements beyond the university and his ready availability to the popular press not only provided him with useful supplementary income but also gave him wider exposure than a professor would normally enjoy. How many policy-makers ever read the *Annals of the Association of American Geographers* or the *Transactions of the Institute of British Geographers*? Lattimore's insightful analyses of Asian problems made his influence even more enduring than that of the political journalists of his day – e.g., Walter Lippmann and Drew Pearson (the latter a former student of the geographer J. Russell Smith and a particular nemesis of Joseph McCarthy, who called him 'the mouthpiece of international Communism'). It is likely, too, that Lattimore's panoramic views of Old World culture history will endure longer than those of the metahistorians Arnold Toynbee and Oswald Spengler because he had a firmer grasp of historical and geographical reality. Eric Hobsbawm sees Lattimore as a sort of spiritual descendant of Ibn Khaldun in his sweeping construction of a history of Asian societies (*On History*, 1997, p. 93). Lattimore's history was 'down-to-earth', and his geography had a time dimension that was missing from the work of so many academic geographers. He appreciated the cultural and historical geography of the Berkeley School, but he was more than a little sceptical of the work of contemporary urban and economic geographers. As he said to Marvin Mikesell in 1955, 'The state of geography in this country is deplorably low. Counting how many people go into a supermarket rates as "economic geography".' Lattimore's own economic geography would be more concerned with tribal markets, animal domestication, and caravan traffic. In his Paris years he found in Xavier de Planhol a geographer with similar interests. He also had great admiration for French historians of the *Annales* School.

The geographer Theodore Herman, another 'old China hand', has recently given the following description of his friend Owen Lattimore: 'I still have a very strong impression of him as mentally sharp, a striking speaker and writer, with a good sense of humor, modest and quite reserved in a group, and very considerate of others but with little time for fools ... I would rank him intellectually with the French regional geography-history school that expresses a multi-discipline understanding of human society. Like Carl Sauer, though without the formal training, OL's work is always stimulating to those who favor an integrated approach.' Professor Herman sees Lattimore's legacy in the following themes: '1. The political and economic importance throughout history of the climatic divide between the settled and the nomadic societies of inner Asia recognized by building the Great Wall; 2. The spread of permanent Chinese movement into NE China in the 20th century, one of the last large-scale farming frontiers of the earth; and 3. Familiarizing us with the possible revival in importance of the ancient cities along the Silk Route' (letter from T. Herman, 13 May 1997).

Geographers and scholars in related disciplines have continued to return to Lattimore's seminal works, *Inner Asian Frontiers of China* in particular, to refresh themselves at the wellsprings of his thought. Arthur Waldron has produced a definitive study of the Great Wall(s) of China (*The Great Wall of China: From History to Myth*, 1990), a topic that continued to fascinate Owen Lattimore throughout his entire career. Piper Gaubatz (*Beyond the Great Wall*, 1996) has found much of value in Lattimore's ideas and has extended them even further into areas on which he spent little time, most notably frontier cities and Tibet. In venturing beyond the bounds of a single academic discipline, other contemporary scholars, such as Wolfram Eberhard, Edward Schafer and Paul Wheatley, have shared Lattimore's vision and eclecticism.

The greater the length of one's bibliography and the greater one's public exposure, the longer the gauntlet that one has to run. Living in the midst of the Asian wars, Lattimore was in a position to report on matters of special importance to the world, but he was also made vulnerable to attacks from distant viewers who were unhappy with his version of the truth because it did not fit their preconceived notions. A man of firm and consistent beliefs, he was nevertheless able to change his mind and even to ridicule some of his own previously published statements, a rare ability unfortunately not shared by his politically conservative critics. His name continues to be invoked, not only in discussions of Chinese and Mongolian affairs but also in the struggle for academic freedom in America.

Acknowledgements

For assistance with this project, I should like to thank Roy Goodman and the staff of the American Philosophical Society Library and Manuscripts Division; the staff of the Manuscript Division of the Library of Congress; James Stimpert and staff of the Ferdinand Hamburger, Jr Archives of The Johns Hopkins University; Anne Ostendarp, Archivist, Dartmouth College Library; Professor C. Julian Bishko of the University of Virginia; Professor Theodore Herman of Colgate University; Sir Laurence Kirwan of the Royal Geographical Society; Professor David Lattimore of Brown University; Professor W.R. Mead of University College London; Professor Marvin W. Mikesell of the University of Chicago; Professor Robert P. Newman of the University of Iowa; and Professor M. Gordon Wolman of The Johns Hopkins University.

Bibliography and Sources

1. ARCHIVAL SOURCES

The following repositories were the sources of the manuscript materials used in writing this essay: American Philosophical Society, Philadelphia; Library of Congress, Manuscript Division, Washington, DC; National Geographic Society, Washington, DC; The Johns Hopkins University, Ferdinand Hamburger, Jr Archives, Baltimore; and Dartmouth College Archives, Hanover, New Hampshire.

2. A SELECTION OF BIOGRAPHICAL SOURCES ON OWEN LATTIMORE

Boas, George, and Wheeler, Harvey (eds), *Lattimore the Scholar*, the editors, Baltimore, 1953. 61 pp.

Cotton, James, *Asian Frontier Nationalism: Owen Lattimore and the American Policy Debate* (Studies on East Asia), Humanities Press International, Inc., Atlantic Highlands, New Jersey, 1989. 181 pp.

Halkovic, Stephen A., 'Professor Owen Lattimore: a bibliography [1920–1972]', in *Analecta Mongolica*, Publications of the Mongolia Society, Occasional Papers, No. 8, The Mongolia Society, Inc., Bloomington, Indiana, 1972, 123–42.

Hangin, John G. and Onon, Urgunge (eds), 'Professor Owen Lattimore: a biographical sketch', *Analecta Mongolica*, 1972, 10–18.

Lattimore, David, 'Biographical note', in Eleanor Holgate Lattimore, *Turkestan Reunion*, Kodansha America, Inc., New York, 1994, 319–23. (Reprint of book originally published in 1934; includes Owen Lattimore's Introduction to the 1975 edition, with a new foreword by Evelyn Stefansson Nef.)

Lattimore, David, 'Introduction', in Owen Lattimore, *The Desert Road to Turkestan*, Kodansha America, Inc., New York, 1995, ix–xxiv, 373–4. (Reprint of book originally published in 1928; includes Owen Lattimore's Introduction to the 1975 edition.)

Lattimore, Owen, *China Memoirs: Chiang Kai-shek and the War Against Japan*, compiled by Fujiko Isono, University of Tokyo Press, Tokyo, 1990. xi + 252 pp. An autobiographical work, mostly made up from taped interviews with Fujiko Isono in the years 1982–4 and edited by John DeFrancis.

Lattimore, Owen, 'Happiness Is Among Strangers' (autobiography), 140-page typescript, written between 1977 and 1980, covering OL's youth to about 1925, with asides referring to later events. Considered by Oxford University Press but never published. Copy in Lattimore Papers in Library of Congress was missing first 85 pages (except for p. 80), but in 1997 Professor Newman gave the Library of Congress a copy that was almost complete.

Lattimore, Owen, 'Preface', in *Studies in Frontier History*, Oxford University Press, London, 1962, 11-31.

Lewis, Lionel S., *The Cold War and Academic Governance: The Lattimore Case at Johns Hopkins*, State University of New York Press, Albany, 1993. xii + 318 pp.

Newman, Robert P., *Owen Lattimore and the 'Loss' of China*, University of California Press, Berkeley, Los Angeles, and Oxford, 1992. xvi + 669 pp.

Schell, Orville, 'Ordeal by slander: Owen Lattimore's odyssey from Mongolia to Washington', in Owen Lattimore, *High Tartary*, Kodansha America, Inc., New York, 1994, ix–xxx, 377–8. (Reprint of book originally published in 1930; includes Lattimore's Introduction to the 1975 edition.)

3. BOOKS BY OWEN LATTIMORE

1928   *The Desert Road to Turkestan*, Methuen and Co., Ltd, London. xiv + 331 pp. Published in Boston by Little, Brown, and Company, 1929. (Reprinted by AMS Press, New York, 1975, with new eight-page Introduction by Owen Lattimore (dated 1972). Reprinted by Kodansha America, Inc., New York, 1995, with new sixteen-page Introduction by David Lattimore.)

1930   *High Tartary*, Little, Brown, and Company, Boston. xvii + 370 pp. (Reprinted by AMS Press, New York, 1975, with new five-page introduction by Owen Lattimore (dated 1973). Reprinted by Kodansha America, Inc., New York, 1994, with new 22-page Introduction by Orville Schell.)

1932   *Manchuria, Cradle of Conflict*, The Macmillan Company, New York. xvii + 311 pp. (Revised edition 1935.)

1933   *The Gold Tribe: 'Fishskin Tatars' of the Lower Sungari*, Memoirs of the American Anthropological Association, No. 40. 77 pp.

1934   *The Mongols of Manchuria*, George Allen & Unwin, Ltd, London. 311 pp. (Published in New York by The John Day Company.)

1940   *Inner Asian Frontiers of China*, American Geographical Society, Research Series, No. 21, American Geographical Society, New York. xxiii + 585 pp. (Second edition, Capitol Publishing Company, Inc., Irvington-on-Hudson, NY, and the American Geographical Society, 1951, with new introduction by OL, xvii–xlviii. Paperback of second edition published by Beacon Press, Boston, 1962. Second edition, with new Introduction by Alastair Lamb, published by Oxford University Press, Hong Kong, 1988.)

1941   *Mongol Journeys*, Doubleday, Doran and Co., Inc., New York. x + 324 pp. (Published in London by J. Cape.)

1944   (with Eleanor Lattimore) *The Making of Modern China: A Short History*, W.W. Norton & Company, Inc., New York. 212 pp. (Revised version issued as *China: A Short History*, 1947.)

1945   *Solution in Asia*, Little, Brown and Company, Boston. 214 pp. (Published in London by the Cresset Press.)

1949   *The Situation in Asia*, Little, Brown and Company, Boston. 244 pp.

1950   *Ordeal by Slander*, Little, Brown and Company, Boston. viii + 236 pp. (Published in London by MacGibbon and Kee, 1952.)

1950   *Pivot of Asia: Sinkiang and the Inner Asian Frontiers of China and Russia*, Little, Brown and Company, Boston. xiv + 288 pp.

1962   *Studies in Frontier History: Collected Papers, 1928–1958*, Oxford University Press, London. 565 pp.

1962   *Nomads and Commissars: Mongolia Revisited*, Oxford University Press, New York. xxiii + 238 pp.

1968   (with Eleanor Lattimore) *Silks, Spices and Empire: Asia Seen Through the Eyes of Its Discoverers* ('The Great Explorers Series', ed. Evelyn Stefansson Nef), Delacorte Press, New York. xi + 340 pp.

4. A SELECTION OF LATTIMORE'S PUBLISHED ARTICLES

Owen Lattimore was a prolific author of newspaper and magazine articles. For fuller bibliographies, the reader should consult the list published by Stephen Halkovic in *Analecta Mongolica*, cited above in section 2, and the bibliographies at the end of Lattimore's *Studies in Frontier History* (section 3) and Newman's biography (section 2). Below are listed several articles that are of interest to geographers. The papers that were reprinted in *Studies in Frontier History* are marked [SFH].

1928   'Caravan routes of Inner Asia', *Geographical Journal*, Vol. 72, No. 6 (December), 497–531. [SFH]

1929   'The Desert Road to Turkestan', *National Geographic Magazine*, Vol. 55, No. 6 (June), 661–702.

1932   'Byroads and backwoods of Manchuria', *National Geographic Magazine*, Vol. 61, No. 1 (January), 101–30.

1932   'Chinese colonization in Manchuria', *Geographical Review*, Vol. 22, No. 2 (April), 177–95. [SFH]

1932   'Chinese colonization in Inner Mongolia: its history and present development', in W.L.G. Joerg (ed.), *Pioneer Settlement*, American Geographical Society, New York, 288–312.

1933   'The unknown frontier of Manchuria', *Foreign Affairs*, Vol. 11, No. 2 (January), 315–30.

1934   'A ruined Nestorian city in Inner Mongolia', *Geographical Journal*, Vol. 84, No. 6 (December), 481–97. [SFH]

1935   'Caravan travel in Mongolia', in George C. Shattuck (ed.), *Handbook of Travel*, Harvard University Press, Cambridge, Massachusetts, 63–76.

1937   'The mainsprings of Asiatic migration', in Isaiah Bowman (ed.), *Limits of Land Settlement*, Council on Foreign Relations, New York, 119–35. [SFH]

1937   'Origins of the Great Wall of China: a frontier concept in theory and practice', *Geographical Review*, Vol. 27, No. 4 (October), 529–49. [SFH]

1938   'The geographical factor in Mongol history', *Geographical Journal*, Vol. 91, No. 1 (January), 1–20. [SFH]

1938   'Descendants of Genghis Khan, II: Mongols of the Chinese border', *Geographical Magazine*, Vol. 6, No. 5 (March), 327–44.

1942   'China opens her wild west', *National Geographic Magazine*, Vol. 82, No. 3 (September), 337–67.

1943   'Yunnan, pivot of Southeast Asia', *Foreign Affairs*, Vol. 21, No. 3 (April) 476–93.

1944	'New road to Asia', *National Geographic Magazine*, Vol. 86, No. 6 (December), 641–76.
1944	'The inland crossroads of Asia', in H.W. Weigert and V. Stefansson (eds), *Compass of the World*, The Macmillan Company, New York, 374–94. [SFH]
1946	'The Outer Mongolian horizon', *Foreign Affairs*, Vol. 24, No. 4 (July), 648–60. [SFH]
1947	'An Inner Asian approach to the historical geography of China', *Geographical Journal*, Vol. 90, Nos 4–6 (October–December), 180–7. [SFH]
1949	'Mongolia: filter or floodgate?', *Geographical Magazine*, Vol. 22, No. 6 (October), 212–21.
1949	'Inner Asian frontiers', in H. Weigert, V. Stefansson and R.E. Harrison (eds), *New Compass of the World*, Macmillan, New York, 262–95. (Previously published in *Journal of Economic History*, Vol. 7, No. 1 (May 1947), 24–52.) [SFH]
1949	'Yakutia and the future of the North', in H. Weigert, V. Stefansson and R.E. Harrison (eds), *New Compass of the World*, Macmillan, New York, 135–49. [SFH]
1953	'The new political geography of Inner Asia', *Geographical Journal*, Vol. 119, Part 1 (March), 17–32. [SFH]
1955	'The frontier in history', in *Relazioni del X Congresso Internazionale di Scienze Storiche* (Rome), Vol. 1 (September), 103–38. [SFH]
1961	'The social history of Mongol nomadism', in W.G. Beasley and E.G. Pulleyblank (eds), *Historians of China and Japan*, Oxford University Press, London, 328–43.
1963	'The geography of Chingis Khan', *Geographical Journal*, Vol. 129, Part 1 (March), 1–7.
1963	'Chinghis Khan and the Mongol conquests', *Scientific American*, Vol. 209, No. 8 (August), 54–60, 62, 64, 66, 68.
1973	'Return to China's northern frontier', *Geographical Journal*, Vol. 139, Part 2 (June), 233-42.
1979	'Herdsmen, farmers, urban culture', in l'Equipe écologie et anthropologie des sociétés pastorales (ed.), *Pastoral Production and Society/Production pastorale et société* (Proceedings of the International Meeting on Nomadic Pastoralism, Paris, 1–3 December 1976), Cambridge University Press, Cambridge, and Editions de la Maison des Sciences de l'Homme, Paris, 479–90.
1980	'The periphery as locus of innovation', in Jean Gottmann (ed.), *Centre and Periphery: Spatial Variation in Politics*, Sage Publications, Beverly Hills, CA and London, 205–8.

Chronology

1900	Born 29 July in Washington, DC
1901–21	Family lived in China, where David Lattimore was professor of English in Shanghai (1901–5), Paotingfu (1905–13) and Tientsin (1913–21)
1912–19	Attended school in Switzerland (1912–14) and England (1914–19)
1919–26	Worked for various commercial enterprises in Tientsin, Shanghai and Peking
1926	Married Eleanor Holgate, 4 March
1928	*The Desert Road to Turkestan*
1928–9	Postgraduate student in anthropology, Harvard University
1934–41	Editor of *Pacific Affairs*
1938–63	Lecturer, then Professor, The Johns Hopkins University; Director, Walter Hines Page School of International Relations, 1939–55
1940	*Inner Asian Frontiers of China*
1941–2	Political adviser to Chiang Kai-shek
1942	Awarded Patron's Medal of the Royal Geographical Society
1942–4	Director, Pacific Bureau of Office of War Information, San Francisco
1950–5	Brought before committees of the US Senate to testify about alleged Communist ties, 1950 and 1952; indicted for perjury 1952 and 1954; all charges dropped 1955
1962	*Studies in Frontier History*
1963–70	Professor and Head, Department of Chinese Studies, University of Leeds
1970	Eleanor Lattimore died 21 March
1986	Recipient of Honors Award from the Association of American Geographers
1989	Died 31 May in Providence, Rhode Island

Gary S. Dunbar is Professor (Emeritus) of Geography in the University of California, Los Angeles. His address is 13 Church Street, Cooperstown, New York 13326, USA.

Tsunesaburo Makiguchi

1871–1944

Keiichi Takeuchi

Since ancient times in Japan, as in all civilizations, a system of knowledge has existed pertaining to the environment and places where humans lived, and other places to which they were connected through trading or other activities. Also to be found is a tradition of compilation of geographical information on territories carried out by the rulers of these territories for administrative purposes. The present-day Japanese term *chiri*, corresponding to the Western term 'geography' and consisting of a combination of two ideograms borrowed from the Chinese classics, has, however, no relationship to the above-mentioned indigenous tradition of geographical knowledge and geographical description, it being an invention by Japanese specialists in Western studies at the beginning of the nineteenth century for the purpose of rendering the Western term 'geography' into Japanese.

Nowadays, mainly because of the institutionalization of geography as a serious study subject in the school curriculum, the term *chiri* is used as a matter of course in daily life. In fact, after the promulgation of the compulsory education system in 1872, geography came to be considered, along with Japanese and history, as a significant means of inculcating the people with nationalist sentiments. With geography now being a required subject, people were needed to teach it, and the study of geography and how to teach it became mandatory at teacher training colleges and subsequently at the Higher Normal schools. However, those who taught geography at secondary and higher educational institutions up until the end of the nineteenth century cannot be considered specialists in geography; for teaching material they relied on sporadic translations of or adaptations from Western geographical books, and they gave lectures whose texts were based on Western textbooks of geography. The examination for the teacher's licence for the teaching of geography in secondary schools was set up in order to cope with the perennial shortage of geography teachers in secondary schools; applicants undertaking the examination were primary school teachers. But the examiners themselves continued to be the same kind of non-academicians who fell back on second-hand geographical material when teaching. It was not until the 1890s that a certain number of specialist geographers who had studied abroad or were seriously engaged in geographical research began to teach at higher educational institutions, including the imperial universities. Thus the institutionalization of geography as a research subject may be said to have taken place at the end of the nineteenth century or the beginning of the twentieth, becoming stabilized with the founding of the first Department of Geography at the Imperial University of Kyoto in 1907 (Noma, 1976). During the approximate thirty-year time lag between the adoption of geography as a school subject in the national education system and the institutionalization of geography as a scientific research discipline, a certain number of figures were to be found working in the field of geography. Tsunesaburo Makiguchi was one of these interesting and individualistic people. Whilst still a young man Makiguchi became a geography teacher at the Teacher Training School in Hokkaido, and his first book, *Jinsei chirigaku* (*Geography of Human Life*), was published in 1903, actually preceding the establishment of academic geography in Japan (Takeuchi and Nozawa, 1988).

The second aspect for which Makiguchi is noted is his devotion to Buddhism. Where the general public is concerned, the geographical aspects of his work and his related accomplishments are largely eclipsed by his fame as the founder of the ubiquitous and powerful Buddhist association Soka Gakkai. His adoption in the late 1920s of the faith of the Orthodox Nichiren sect of Buddhism can in part be explained as having been due to the troubled social and political situation of Japan at that time, in which he met with difficulty in promoting his movement involving school education; Buddhistic teachings afforded a better vehicle, he found, for the spreading of his ideas. In any case, while he was later to further emphasize the supremacy of the teachings of the Lotus Sutra, he never ceased to maintain his profound interest in and systematic pursuit of geography that began in his student years.

1. Education, Life and Work

Tsunesaburo Makiguchi was born in 1871 in a desolate fishermen's village on the coast of the Japan Sea, in Niigata prefecture. He was then named Choshichi Watanabe. Upon the divorce of his parents in 1877, he was adopted by a distant relative, Zentayu Makiguchi, operator of a small shipping agency, and thus acquired the name by which he is now known. In 1885, he moved to Hokkaido where an uncle helped him find employment as a factotum at the Otaru police station; at the same time he continued his studies, and in 1893 graduated from the Teacher Training School of Hokkaido. Because of his brilliant scholastic record he was subsequently engaged as teacher at the elementary school attached to the Teacher Training School. From his student days he had always been deeply interested in geography, and in 1896, at the age of 25 he passed the examination for the teacher's licence for the teaching of geography in secondary schools. At that period (1885–1947), teachers of secondary schools were recruited from among graduates of universities and Higher Normal schools. In the case of geography, because of the lack of geography courses at university level, secondary school geography teachers were partly recruited from among primary school teachers who had passed the above examination (Sato, 1988). There was always a large number of aspirants eager to take this examination, which generally took a number of successive attempts, over several years, to pass. Makiguchi's success in passing at so early an age was therefore unusual and indicative of his exceptional ability. In 1897, he was appointed teacher of geography at the Hokkaido school from which he had graduated. In 1901, however, he resigned this post and left for Tokyo.

The circumstances of his resignation are not clear. Some researchers suggest that his sympathy with the dissident students protesting against the militarist trends in education as practised at the school prompted him to resign in a gesture that also expressed his own personal disagreement with the educational policy (Isonokami, 1993). In studying geography, moreover, he felt the lack of a systematic treatise on human geography in Japan; upon arriving in Tokyo, he devoted himself to the task of writing one. Consequently, in 1903, he paid a visit to Shigetaka Shiga, who was then teaching geography at Tokyo Semmon Gakko (now Waseda University). Shiga graduated from Sapporo Agronomical College in 1884 in its early years when all the professors were either Americans or Englishmen, who inspired the students to an interest in the earth sciences in the broader sense through the teaching of agronomical sciences. Among the graduates of the early years of the college were Kanzo Uchimura, author of an outline of geography published in 1894, which was actually an adaptation of A. Guyot's book *The Earth and Man* (1849), and Inazo Nitobe, who later introduced A. Meitzen's settlement studies into Japan. Nitobe, together with Kunio Yanagita, encouraged the pursuit of Japanese folklore studies, in connection with which Makiguchi later attended meetings. Shiga's *Nihon fukeiron* (*Japanese Landscape*) was published in 1894. This work became famous for its exaltation of the proper scenic beauty of Japan and its author was considered a connoisseur of recent achievements in Western geography. Makiguchi, having resigned from his teaching job and being then totally unknown, asked Shiga to revise his manuscript, which had reached voluminous proportions. Complying with Shiga's suggestions, he shortened the manuscript to about half its original length and published the results as *Jinsei chirigaku* (*Geography of Human Life*) with an introduction by Shiga. The first edition consisted of more than a thousand pages. At that time, some short treatises on geography by Shiga and Uchimura had already been published. Makiguchi's book was, however, considerably more voluminous, comprehensive and systematic, covering, as it did, all fields of human geography. This book, up until its tenth edition in 1909, was the most read of the reference works consulted by the aspirants to the examination for the teacher's licence for the teaching of geography in secondary schools.

In Tokyo, in order to support himself and his family, Makiguchi retained various jobs, including editing and part-time lecturing. It was not until 1909 that his appointment to the post of head of the teaching staff of an elementary school in Tokyo provided him with a steady source of income. After 1910, for some years he worked part-time at the Ministry of Education, where he was engaged in the compilation of geography textbooks and atlases for the elementary school level. In 1912, his book *Kyoju no togo chushin to shite no kyodoka kenkyu* (*Considerations on Homeland Studies as the Integrating Focus of School Education*) was published. Based on his experiences as a geography teacher, it was also meant to serve as a practical application of the teachings contained in his *Geography of Human Life*. Subsequently, in 1916, a further work, *Chiri kyoju no hoho oyobi naiyo no kenkyu* (*Studies on Methods and Content in the Teaching of Geography*) also saw publication.

Life in Tokyo afforded many new opportunities enabling him both to broaden and to deepen the scope of his geographical studies. In 1909, for instance, he made the acquaintance of Kunio Yanagita, founder of the so-called Japanese folklore school, and hence had opportunities to conduct numerous field studies with him in various rural parts of the country, often under the sponsorship of the Ministry of Agriculture and Commerce, where Yanagita served as a senior official. In 1910, Yanagita commenced the holding of regular study-meetings called *kyodo-kai* (literally, 'group or association for homeland studies') at the house of Inazo Nitobe. Makiguchi, along with a number of academic geographers such as Michitoshi Odauchi, was a member of the *kyodo-kai*, and this organization exercised a considerable influence on the formation of early academic geography in Japan (Takeuchi, 1984–5).

In 1912, he met a young teacher, Jogai Toda, who later became his close collaborator within the association he founded in 1930, Soka Kyoiku Gakkai, literally 'Society for Value-creating Pedagogy', which after World War II was to become today's militant Buddhist association Soka Gakkai, and Toda its second President (though its official first President, Makiguchi, died as President of the association when it was still Soka Kyoiku Gakkai). Where this aspect of Makiguchi's career is concerned, it is necessary to examine the circumstances in which he became a follower of the Nichiren Shoshu or Orthodox Nichiren sect from which the teachings embodied in Soka Kyoiku Gakkai were derived. From 1918 he worked as headmaster of night schools located in poverty-stricken districts and specializing in courses at the elementary level. In the course of this work he emphasized what he had felt since his days as a teacher at Hokkaido, that the purpose of education was to provide practical knowledge useful for social life, and also stressed the necessity of providing an education suitable to each stage of the development of children. This was in accordance with the Pestalozzian educational principles he had absorbed from Takeshi Shirai, his colleague when he was a teacher at the Teacher Training School in Hokkaido (Nakagawa, 1981). The social activities of Makiguchi as headmaster of the night schools met with the antagonism of influential local persons, who instigated a movement to expel him from his position. But he managed to withstand these onslaughts and maintain his position thanks to the support of scholars in homeland studies, and other influential scholars such as Tamon Maeda.

Japanese society in the 1920s had fallen upon hard times: depression following upon World War I, the disastrous earthquake which struck Tokyo in 1923, the social instability revealed by repeated rice riots and labour disputes, and so on. Around the middle of the 1920s Makiguchi conceived and began to write the synthesis of his educational philosophy and practice under the name of 'value-creating education'. He had been born into a family following the Zen sect, while the Makiguchi family into which he was adopted belonged to the Tendai sect of Buddhism; then during his youth he was considerably influenced by Protestantism through Kanzo Uchimura and Inazo Nitobe. Nevertheless, he never actually belonged to any particular religious sect. It was only in 1928, when he met Motohira Mitani, headmaster of Mejiro Commercial School, that he was introduced to and became a follower of the Orthodox Nichiren sect.

In 1260, Nichiren, the founder of the sect named after him, wrote *Rissho ankokuron* (*On Securing the Peace of the Land Through the Propagation of True Buddhism*), based on the teachings of the Lotus Sutra (one of the Mahayana sutras; in philosophical terms, a Buddhist teaching leading people to enlightenment). At this time, the situation of Japanese society was a chaotic one, menaced as it was by a Mongol invasion as well as being disrupted by internal revolt. In this work Nichiren insisted on the necessity of following the doctrine of the Lotus Sutra in order to realize social justice and establish social order.

Makiguchi introduced Toda to Mitani, soon after which Toda also became a follower of the Orthodox Nichiren sect. After that, the activities Makiguchi pursued in collaboration with Toda, while continuing to constitute a reformist educational movement, at the same time developed into a fanatical religious movement based on an original interpretation of the Nichiren doctrine. Where Makiguchi was concerned, his devotion to the Nichiren orthodoxy, with its premises for the realization of social justice, was a continuation and perfecting of his geographical research and pursuit of educational ideals. But it is necessary to note the difficulties and social contradictions Makiguchi faced in the circumstances of his conversion to the Nichiren faith.

The first volume of *Soka kyoikugaku taikei* (*System of Value-creating Pedagogy*) was published in 1930 and followed by successive volumes, the fourth appearing in 1934. From chapter 5 of Volume 2, numerous references to the Lotus Sutra began to appear. In all four volumes, the current social and political situations in Japan are often mentioned and a great deal of the literature referred to consists of publications dating from 1920 onwards. We gather from these facts that this four-volume work was very firmly rooted in the Japanese reality of the time, and that Makiguchi accomplished this task at the same time as fulfilling his duties as school headmaster. The activities of Makiguchi and Toda encountered the opposition of political circles of ultra-nationalist and State Shintoist persuasions, and in 1932, at the age of 61, Makiguchi was compelled to resign from the last post he held, that of elementary school headmaster. Increasing emphasis had come to be laid on a Tennoism based on a nationalistic form of Shintoism which, particularly in the 1940s, held the Emperor (*tenno*) and the imperial family to be direct descendants of the supreme deity, Amaterasu O-Mikami (the Sun Goddess in Japanese mythology). After 1942, the ultra-nationalist regime ordered all religious sects in Japan to enshrine in their places of worship inscribed tablets representing the ancestral deities of the *tenno* family. All the established Buddhist and Christian sects in Japan complied with this order: even the Taiseki temple, head temple of the Orthodox Nichiren sect at Fujinomiya in Shizuoka prefecture, followed suit, and in June 1943, summoned Makiguchi, Toda and other leaders of their group to the temple, where a tablet was to be allotted to them as prescribed. Makiguchi, simply on the grounds that any doctrine other than that of Buddhism was unacceptable, categorically refused to accept the tablet, thus incurring the wrath of the military group and the ultra-nationalist political body (Kumagaya, 1978). A week later, he was arrested on charges of disrespect towards State Shintoism and offences against the Maintenance of Public Order Act. The following year, in common with a number of politicians and other intellectuals who voiced their opposition to the current regime, he died in Tokyo's Sugamo prison, of old age and malnutrition, compounded by grief at the news of the death of his son at the battlefront.

Toda was imprisoned on similar charges, but survived and was released in July 1945, one month before the Japanese surrender and the consequent end of World War II. In 1946, he re-established the association under the name of Soka Gakkai, which ceased to maintain its original educational aspects and became purely religious in nature. It was not until 1951, however, that Toda officially assumed the post of second President. His experiences in prison had evoked in him a spiritual awakening on the basis of which he systematized the greater part of the doctrine proper to Soka Gakkai as it is today. He placed emphasis on the revolutionary transformation of life as lived by human beings (*ningen kakumei*). The second emphasis was on the realm of Buddha within Japan of the Latter Day of the Law, that is, Japan as it now was – hence the intensive proselytizing activities and aggressive involvement in politics. As regards the second emphasis, the development of Makiguchi's Buddhist ideas in the Soka Gakkai movement after World War II is discernible. In November 1948, Soka Gakkai organized a memorial service in Tokyo on the third anniversary of Makiguchi's death, attended by about 700 members of the faith. Five years later, nation-wide membership had risen to 500,000 people.

2. Scientific Ideas and Geographical Thought

The title of Makiguchi's *magnum opus* of 1903, *Jinsei chirigaku* (*Geography of Human Life*) was very novel at that period, and nobody has ever used the term since. Already, in 1889, in the first issue of *Chigaku Zasshi* (*Journal of Geography*), the organ of the Tokyo Geographical Society, Bunjiro Koto, then Professor of Geology at the Imperial University of Tokyo, in his paper on the meaning of geography used the terms *jinrui chirigaku*, i.e., anthropogeography, and *jimmon chirigaku*, i.e., human geography. Makiguchi was perfectly aware of the existence of these terms; in the preface to *Geography of Human Life*, and also in chapter 33 of the first edition, he explained his reasons for and the appropriateness of adopting this particular term for his title. Where education in geography was concerned, the necessity of integrating fragmented regional descriptions was suitably underlined by the term 'geography of human life'. For Makiguchi, this integration or systematization hinged on the relationship between humans and the physical environment, explained from the viewpoint of human activities or human experiences.

> Geography constitutes a systematic knowledge of the relationship of physical phenomena and phenomena concerning human lives, both of which are distributed over the earth's surface with a certain regularity. In short, geography is a science which explains the relationship between earth and human life. (Makiguchi, 1903, p. 988)

The key concept of *Geography of Human Life* was the importance of geography in school education and the role of human activities or human experiences in the explanation of the relationship between humanity and nature. It is very understandable that Makiguchi adopted the humanity–nature paradigm in geography, as in nineteenth-century geography, both pre- and post-Darwin, this paradigm clearly predominated. In fact, in chapter 33, he cited Carl Ritter when defining geography, and moreover emphasized the fact that the uniqueness of geography consisted in the systematic and theoretical explanation of the relationship from the viewpoint of daily practices in the life of human beings. Numerous discourses common to both Makiguchi's geography and present-day humanistic geography are to be found. At the same time, his anthropocentric understanding of the humanity–nature relationship prepared the way for the further development of his utilitarianism, which was later to be formulated more explicitly.

It should be pointed out that academic geographers have never understood the significance of Makiguchi's term *jinsei chirigaku*. Immediately after the book appeared, Takuji Ogawa's review of it, which appeared in the *Chigaku Zasshi* of 1904, criticized the use of this term because the contents of the work in question in fact consisted of cultural or human geography, and moreover the term *jinsei chirigaku* itself seemed inordinately strange. In 1978, Hisaya Kunimatsu, graduate in geography of the Imperial University of Tokyo and a prominent academic geographer, published a book (Kunimatsu, 1978) on *Geography of Human Life* which constituted the very first study written by an academic geographer on Makiguchi's geography. In his work, Kunimatsu recognized Makiguchi's acceptance of the concept of the nineteenth-century paradigm of geography realized through the reading of translations of Western geographical writings, and recognized, moreover, that the leitmotif of *Geography of Human Life* was the importance of geographical education. But at no time did he understand that Makiguchi had been greatly influenced, since his period as a young teacher of geography, by the

Pestalozzian method of direct observation. Only from the Pestalozzian viewpoint applied to geographical education can we properly comprehend Makiguchi's reasons for using the term *jinsei chirigaku*. The legacy of Pestalozzian methods was especially strong around the end of the nineteenth century at teacher training schools in Nagano and Hokkaido (Nakagawa, 1978).

Part 3 of *Geography of Human Life*, which contained a systematic and comprehensive treatment of the economic and political activities of human beings on the earth's surface, can rightly be considered the most important section of the book. In the eighth edition, published in 1908, Makiguchi added one chapter dealing with the sentimental attachment of the social group to locality, the place, that is, where they were born and/or where they lived. This new chapter was rather short, but it is to be noted that with its addition, Makiguchi's geography took on a more humanistic orientation. We have also to note that from the first edition, Part 3 also constituted a highly systematic treatise on economic geography and settlement geography. Makiguchi very rarely read foreign literature in the original languages and we are compelled to admire the thoroughness of his knowledge of agriculture and industrial location theories acquired almost solely through the reading of translations of Western works. He did not cite the name of von Thünen, but he explained, albeit with some errors of understanding, the theory of isolated state, something that was rather exceptional even in Western geographical writings of that period (Ohji, 1982). In his explanation of industrial location, Makiguchi particularly noted the complicated relationship between transportation costs of raw materials and of products.

Makiguchi wrote two books on geographical education which were published in 1912 and 1916. He prepared more fully-developed ideas on geographical education with the intention of having them published as a volume of his series entitled *Soka kyoikugaku taikei* (literally, *System of Value-creating Pedagogy*). We cannot know what and how much he wrote for this volume, however, as the manuscript was confiscated by the police at the time of his arrest in 1943 and has never been found.

Considerations on Homeland Studies as the Integrating Focus of School Education of 1912 was a clear manifestation of the emphasis he placed on homeland studies and on the reconstruction or reorganization of the school curriculum based on direct observation, along Pestalozzian principles, of the environment of the pupils. Already in 1890, the Imperial Decree on Education, which emphasized a centralized nationalistic type of school education, had been promulgated, and Makiguchi's position was challenged by this line of thinking. His book, meantime, reached its tenth edition by 1933, and in the 1930s, when the Ministry of Education recommended the pursuit of homeland studies in the framework of a nationalistic reappraisal of the national territory, Makiguchi flatly stated in the preface to the tenth edition of his book that his homeland studies had nothing whatsoever in common with the brand of homeland studies recommended by the Ministry of Education. In the book, he categorically expressed his opposition to the dominant trends in school education, which persisted in maintaining that the supreme purpose of school education was the formation of people as imperial subjects devoted heart and soul to the Emperor. When Makiguchi declared that the target of school education was the creation of the nation, he meant this in the sense of people as citizens rather than subjects.

> About 500 officials belonging to various large and small institutions and functioning in accordance with their given tasks are constantly working with the purpose of increasing the happiness of the people. Thanks to this kind of protection and encouragement [on the part of] the state and to the existence

> of the state powers, our property and lives are safe from thieves and our honour and rights are protected from offences [against them]; even the state powers can intervene [in our affairs] only up to a certain limit. Accordingly, we have to teach schoolchildren the meaning of the state and the purpose of [living] collective lives, and in this way, the minds of schoolchildren will be imbued with a patriotism which is never abstract and never separated as heretofore from the love of their homeland. (Makiguchi, 1912; 1981 edition, Vol. 3, p. 323)

According to the dominant idea, geographical education had to aim, first of all, at the formation of identity with the national territory through the learning of geography. But Makiguchi strongly affirmed that homeland studies, as he saw them, were based on a pedagogical philosophy which held that school education should primarily be useful, where the lives of the children or their happiness were concerned. This sort of utilitarianism was already implicit in *Geography of Human Life* and was stated more explicitly in Makiguchi's writings during the 1930s, after he became a follower of Nichiren Shoshu.

Studies on Methods and Content in the Teaching of Geography of 1916 differed somewhat in character, the larger part of this work being based on lectures he gave at retraining sessions for teachers, in his capacity as veteran teacher and headmaster. His fundamental attitude to school education, however, remained unchanged from that of his previous book: that the purpose of school education was the formation of the nation (*kokumin*, in the sense of 'citizens'; he never used the term 'subject' *shinmin*), and that the study of geography as part of the school curriculum was important in attaining this fundamental purpose. In order to stress the need to increase teaching hours of geography in schools, he also pointed out the importance of geography in daily and professional life in a society where relationships, direct and indirect, with remote parts of the world were expanding. The contents of the book are chiefly given over to the practical knowledge and techniques of geography education in the frame of the school curriculum imposed by the Ministry of Education; nonetheless, a certain amount of discussion proper to Makiguchi appears, such as the stressing of homeland studies as a preparatory stage of geography education in elementary school. He also proposed a fundamental reform regarding the teaching of geography whereby, instead of textbooks, the atlas should be adopted as the principal teaching instrument rather than as secondary material.

> Geography textbooks should be used as complementary material in reading and understanding maps. Only by reversing the relationship between textbooks and atlases in this way, can education in geography manifest its true value. (Makiguchi, 1916; 1981 edition, p. 192)

This was a revolutionary and very idealistic proposal, but at this point it should be remembered that Makiguchi was employed for a while from 1910 at the Ministry of Education, where he worked on the compilation of geography textbooks and atlases as secondary material. When he began his stint at the Ministry, however, editorial work on the second version of geography textbooks for elementary school and on the atlas was already finished and approved by the Ministry. This version of the textbooks continued to be used, with very little to show in the way of modification, up to 1916 and the same applied to the atlas until 1919. It is conceivable that Makiguchi, restricted as he was from being as innovative as he wished concerning the contents of the textbooks, wrote his 1916 book as a form of protest against state authorship of the textbooks of that time.

It should be pointed out here that, at the same time, Makiguchi was by no means completely unaffected by the thinking of the time or the circumstances in which Japan found itself at that period. Hence he not only incorporated the geography of Korea and Formosa into the geography of Japan in chapter 35 of his book, entitled 'Geography education in countries under the influence of the Japanese Empire', but also unequivocally regarded as legitimate the Japanese hegemony over the northeastern part of China, the Russian Far East, Mongolia and the Pacific islands comprising a former German colony. He clearly recognized the Japanese Empire as a multi-ethnic state and in his remarks on the geography of Formosa (Taiwan), whilst recognizing Japanese sovereignty over the island, he wrote with some objectivity that the economic life of the Chinese there was exceedingly competitive in comparison with the indigenous people or the Japanese themselves.

3. Influence and Spread of Ideas

As mentioned above, Tsunesaburo Makiguchi is now well known as the founder of Soka Gakkai, one of the most active and militant Buddhist sects to be found today, and one which also exercises a remarkable influence on the contemporary Japanese political scene.

On the occasion of the publication of the first volume of *Soka kyoikugaku taikei* (*System of Value-creating Pedagogy*), distinguished scholars such as Inazo Nitobe and Kunio Yanagita, his companions in the homeland study group, contributed recommendations, as did a number of politicians holding ministerial posts. The circumstances, however, being entirely different from the time of the publication of *Geography of Human Life*, for which work Makiguchi had asked for and received revision and recommendations from outstanding geographers of that period, there were almost no academicians in pedagogy who welcomed Makiguchi's new work, the few exceptions being a certain select number of teachers and students of pedagogy at the Imperial University of Tokyo. So the influence of *System of Value-creating Pedagogy* among specialists in geography and pedagogy was very limited.

It is necessary that, in order to understand Makiguchi's motive in writing *System of Value-creating Pedagogy*, and also his conversion to the Orthodox Nichiren sect, both writing and conversion be viewed in the social context of Japan at that period, that is, a context made up of the social instability and economic deprivation of the majority of the Japanese people, and the imposition of State Shintoism in the sphere of intellectual life. In the four published volumes, there appeared a large amount of Buddhist terminology he had not hitherto used, but which he now found necessary in order to propound the spiritual ramifications involving his new faith; but his pedagogical thought remained unchanged from that found in his previous writings. The four volumes of *System of Value-creating Pedagogy* comprised the first systematic writings on school education written by a school teacher, composed, furthermore, against the background of the many problems which school education faced in Japan at that time: financial difficulties in schools brought on by the world economic crisis, the increasing burden of military expenditure, increasing state control of and uniformity of school education, as well as inequalities in higher education opportunities. In this sense, where the methodology of school teaching and school management was involved, he was invariably a rationalist and pragmatist and forthrightly criticized the impotence and inability of academic specialists in pedagogy when it came to solving these problems. In the preface to Volume 1, he wrote:

Criticism means nothing to me as I am now – I am losing my sanity, as it were, as I look upon ten million students now facing difficulty in getting into schools, trouble in passing examinations and problems in finding work after graduation.

In chapter 5 of Volume 2, where he treated 'value-creation' in education, for the first time he cited passages from the Lotus Sutra, and from that time onwards continued to do so quite often. In previous writings, Makiguchi's concept of value was a subjective or relative one, or, in terms of economics, he relied on the theory of marginal utility. But observing the perpetual instability and confusion of the Japanese society of that period, he felt the need for an absolute standard of values, and wrote:

One of Nichiren's major writings, *Rissho ankokuron* [*On Securing the Peace of the Land through the Propagation of True Buddhism*], provides an appropriate standard for the appraisal of values in this contemporary world of *mappo* [Latter Day of the Law], beset as it is by the wavering of judgement on the part of people and the uncertainties of life.

Consistently maintaining an anti-ultra-nationalist and anti-militarist stance, in Volume 3 of *System of Value-creating Pedagogy* he actually proposed the strengthening of the autonomy of the school and the adoption of a more open system for the formation of school teachers, a system that would prove operable beyond the confines of the teacher training school system. As stated previously, the utilitarianism of Makiguchi invariably gave priority to the kind of utility that benefited the lives of children and happiness as the ultimate goal of education; and after his conversion to the Orthodox Nichiren sect, the welfare of the people continued to take precedence over all other matters. At the same time, however, he increasingly quoted the Lotus Sutra, stating that man always had to maintain an interaction with his surroundings or environment in order to arrive at a state of harmony with it, finding succour by embracing the teachings of Nichiren. In this way, his utilitarian view with regard to the humanity–nature relationship, as expressed earlier in *Geography of Human Life*, not only remained intact after his conversion to the Orthodox Nichiren sect, but also properly constituted a significant reason for that conversion. He had difficulty propagating his teaching ideas *per se*, under the prevalent social and political oppression, but those ideas were kept alive by being transposed, as it were, into an alternative mode, involving the discipline of the spirit. As it happened, this motivated a growing number of his followers, some of whom were also teachers living out their lives in socially and economically deprived circumstances, to adhere to his teachings.

In conclusion, we can observe, all in all, a certain continuity in Makiguchi's attitude and activities throughout his lifetime, a certain pragmatism or a readiness to realize in concrete form whatever he thought needed to be done. When he was young, he felt the necessity for a treatise of human geography, especially to aid the aspirants to the teacher's licence for the teaching of geography in secondary schools, and so produced in a comparatively short period of time the voluminous *Geography of Human Life*. His second book was published with the practical purposes of promoting and diffusing homeland studies. When he became aware of the critical nature of the situation in school education under the increasing governmental control of the 1930s, he founded Soka Kyoiku Gakkai in order to promote school education having as its primary aim the happiness and well-being of schoolchildren. This pragmatic or utilitarian spirit was unquestionably inherited by the

Soka Gakkai movement after World War II. In fact, among the many Japanese Buddhist sects, which generally attributed more importance to the attaining of Buddhahood and happiness in a Buddhist heaven, only Soka Gakkai gave importance to the realization of Buddhist ideals in this world, for which purpose the political arm of the Soka Gakkai movement was brought into being.

It was not until 1934, six years after his conversion, that in Volume 2 of the *System of Value-creating Pedagogy* he categorically declared it necessary to depend on the Buddhist law rather than human law, and that the ideological basis of his 'value-creating pedagogy' consisted in the teachings of the Lotus Sutra. For Makiguchi, belief in the Lotus Sutra brought about a kind of transcending of human power; nonetheless, in no circumstances did he abandon the rational way of thinking involved in the modern sciences, and it was on the basis of this idiosyncratic logic that he built up and presented a very original interpretation of the Lotus Sutra. His resistance to State Shintoism naturally derived from his attitude, which gave supremacy to the Lotus Sutra, but was at the same time based on his acceptance of the rational logic of the modern sciences.

His death in prison created a certain mystique in the Soka Gakkai movement after World War II; in other words, Makiguchi became a martyr in the invented tradition of the Soka Gakkai. We should note, however, that the Japanese society experienced by Makiguchi was quite different from that in which the Soka Gakkai saw the expansion of its followers to its present enormous numbers. Makiguchi conducted his resistance to the overwhelming power of the establishment of his day. Since World War II, Soka Gakkai has always constituted a reformist movement in the framework of conservative political trends. But at the same time, paradoxically, Makiguchi's teaching, which insisted on the necessity of human endeavour in attaining material and spiritual well-being, found its most fertile soil in the socio-economic condition of the Japan of the rapid economic growth period of the 1950s and 1960s, a period corresponding to the rapid expansion of the Soka Gakkai movement. In that period, many of the common people of Japan held that the more one worked, the more one received, and the more one's standard of living improved. As a result of the expansion and increasing influence of Soka Gakkai in the social, economic and political spheres, the writings of Makiguchi, along with associated commentaries and annotations, have been published in large quantities by Soka Gakkai and related institutions, and the reappraisal of his geographical thought and educational ideals is at least partially due to the success of the Soka Gakkai movement after World War II. Whether or not those followers have genuinely inherited Makiguchi's geographical and pedagogical thought is another issue.

Bibliography and Sources

1. REFERENCES ON TSUNESABURO MAKIGUCHI

Isonokami, G., *Makiguchi Tsunesaburo to Nitobe Inazo (Tsunesaburo Makiguchi and Inazo Nitobe)*, Dai-san Bunmeisha, Tokyo, 1993.

Kumagaya, K., *Makiguchi Tsunesaburo (Tsunesaburo Makiguchi)*, Dai-san Bunmeisha, Tokyo, 1978.

Kunimatsu, H., *Jinsei chirigaku gairon (Outline of Geography of Human Life)*, Dai-san Bunmeisha, Tokyo, 1978.

Nakagawa, K., *Kindai chirikyoiku no genryu* (*Origins of Modern Geography Education*), Kokon Shoin, Tokyo, 1978.

Nakagawa, K., editor's comments on *Kyoju no togo chushin to shite no kyodoka kenkyu* (*Considerations on Homeland Studies as the Integrating Focus of School Education*), Dai-san Bunmeisha, Tokyo, 1981.

Noma, S., 'A history of Japanese geography', in S. Kiuchi (ed.), *Geography in Japan*, University of Tokyo Press, Tokyo, 1976, 3–16.

Ohji, T., 'Waga kuni chirigaku e no Chunen "koritsukoku" no shokai' (Introduction in Japan of von Thünen's 'Isolated State'), in Geographical Institute of Kyoto University (ed.), *Chiri no shiso* (*Geographical Thought*), Geographical Institute of Kyoto University, Kyoto, 1982, 271–88.

Saito, S., 'Jinsei chirigaku kenkyu no tame no joron' (Introduction to the Study of *Geography of Human Life*), in *Complete Works of Tsunesaburo Makiguchi*, Dai-san Bunmeisha, Tokyo, Vol. 2, 1996, 519–43.

Sato, Y., *Senzen no chiri kyoshi. Bunken chirigaku o saguru* (*Geography Teachers in the Pre-war Period. Considerations on the Examination System for the Teacher's Licence for the Teaching of Geography in Secondary Schools*), Kokon Shoin, Tokyo, 1988.

Takeuchi, K., 'Strategies of heterodox researchers in the national schools of geography and their roles in shifting paradigms in geography', *Organon*, Warsaw, Vols 20/21 (1984–5), 277–86.

Takeuchi, K. and Nozawa, H., 'Recent trends in studies on the history of geographical thought in Japan – mainly on the history of Japanese geographical thought', *Geographical Review of Japan*, Vol. 61B (1988), 59–73.

2. SELECTIVE BIBLIOGRAPHY OF WORKS BY TSUNESABURO MAKIGUCHI

1903   *Jinsei chirigaku* (*Geography of Human Life*), Bun'edo Shoten, Tokyo. (Now in Vols 1 and 2 of *Complete Works of Tsunesaburo Makiguchi*, with annotations by Shoji Saito, Dai-san Bunmeisha, Tokyo, 1983, 1996.)

1912   *Kyoju no togo chushin to shite no kyodoka kenkyu* (*Considerations on Homeland Studies as the Integrating Focus of School Education*), Ibunkan, Tokyo. (Now in Vol. 3 of *Complete Works of Tsunesaburo Makiguchi*, with annotations by Hideo Sato, Dai-san Bunmeisha, Tokyo, 1981.)

1916   *Chiri kyoju no hoho oyobi naiyo no kenkyu* (*Studies on Methods and Content in the Teaching of Geography*), Meguro Shoten, Tokyo. (Now in Vol. 4 of *Complete Works of Tsunesaburo Makiguchi*, with annotations by Koichi Nakagawa, Dai-san Bunmeisha, Tokyo, 1981.)

1930–4 *Soka kyoikugaku taikei* (*System of Value-creating Pedagogy*), Vols 1, 2, 3 and 4, (1930, 1931, 1933 and 1934), Soka Kyoiku Gakkai, Tokyo. (Now in Vols 5 and 6 of *Complete Works of Tsunesaburo Makiguchi*, with annotations by Kazunori Kumagaya and Rikio Mokuzen, Dai-san Bunmeisha, Tokyo, 1982, 1983.)

Chronology

1871	Born in Niigata prefecture as Choshichi Watanabe
1877	Divorce of parents. Adopted by a distant relative, Zentayu Makiguchi
1885	Went to Hokkaido. Worked as factotum at the Otaru police station
1893	Changed his name to Tsunesaburo Makiguchi. Graduated from the Teacher Training School of Hokkaido and was appointed teacher at the attached elementary school
1895	Married the second daughter of Zentayu Makiguchi
1896	Passed examination for the teacher's licence for the teaching of geography in secondary schools
1897	Appointed teacher of geography at the Teacher Training School of Hokkaido
1901	Left the Teacher Training School of Hokkaido and went to Tokyo
1903	Visited Shigetaka Shiga, teacher at Tokyo Semmon Gakko, and asked for advice about his forthcoming book. Publication of *Geography of Human Life* in Tokyo
1909	Appointed head of teaching staff of Fujimi Elementary School, Tokyo. Made the acquaintance of Kunio Yanagita and joined in his field researches at Yamanashi prefecture
1910	Joined the *kyodo-kai* founded that year by Yanagita and Inazo Nitobe. Began to work at the Ministry of Education on the compilation of elementary school textbooks
1912	Publication of *Considerations on Homeland Studies as the Integrating Focus of School Education*
1913	Appointed headmaster of Higashimori Elementary School and of Shitaya First Night Course Elementary School
1916	Appointed headmaster of Taisho Elementary School. Publication of *Studies on Methods and Content in the Teaching of Geography*
1918	Began to serve as headmaster of Taisho Night Course Elementary School concurrently with the above
1920	Met Jogai Toda, who became his close collaborator and later President of Soka Gakkai. Appointed headmaster of Mikasa Elementary School. Served concurrently as headmaster of the evening courses school
1922	Appointed headmaster of Shirogane Elementary School
1928	Became a follower of the Orthodox Nichiren sect
1930	Publication of Vol. 1 of *Soka kyoikugaku taikei* (*System of Value-creating Pedagogy*). Founded, in collaboration with Toda, Soka Kyoiku Gakkai (Society of Value-creating Pedagogy), forerunner of Soka Gakkai
1931	Publication of Vol. 2 of *System of Value-creating Pedagogy*. Appointed headmaster of Shimbori Elementary School

1932	Publication of Vol. 3 of *System of Value-creating Pedagogy*. Retired from teaching with the closure of Shimbori Elementary School
1934	Publication of Vol. 4 of *System of Value-creating Pedagogy*
1943	Arrested at Shimoda for refusing to pay allegiance to Tennoist State Shintoism. Toda and other leaders of the movement arrested the same year
1944	Died of old age and malnutrition in Sugamo prison, Tokyo

Keiichi Takeuchi is Professor of Geography at Komazawa University, Tokyo. He was formerly Chairman of the Commission on the History of Geographical Thought of the International Geographical Union.

Thomas Robert Malthus

1766–1834

Patrick H. Armstrong

Courtesy of Haileybury College, Hertford, UK

The year 1998 represented the 200th anniversary of the publication of Malthus' most important work – the first edition of the *Essay on Population* – and so provides an opportune moment for the re-evaluation of his life and contribution. His direct effect on geographical thought was not large; indeed the subject of geography had not emerged as an academic discipline in his lifetime, but the indirect effect, through the influence of his ideas on great areas of thought, and public policy, was substantial. It is for this reason that he is included here. Darling of the modern 'Green' movement, Malthus is also credited with providing Charles Darwin (1809–1882; this series, Vol. 9) with the clue to the mechanism of natural selection that gave his evolutionary theory, set out in *On the Origin of Species* (1859), its coherence (see, for example, Young, 1969). Yet Malthus attracted ridicule and opprobrium in his own time and afterwards:

> He was the 'best-abused man of his age.' Bonaparte himself was not a greater enemy of his species. Here was a man who defended small-pox, slavery and child-murder; who denounced soup-kitchens, early marriage, and parish allowances; who 'had the impudence to marry after preaching against the evils of a family;' who thought the world so badly governed that the best actions do the most harm; who in short, took all the romance out of life and preached a dull sermon on the threadbare text – 'Vanity of vanities, all is vanity.' Such was the character of Malthus as described by his opponents. (Bonar, 1926, p. 1)

1. Education, Life and Work

Thomas Robert Malthus – he preferred the name Robert – was born on 17 February 1766. He came from a scholarly family; his father Daniel (1730–1800) lived amongst his books at the Rookery, near Guildford. Daniel was an acquaintance of Rousseau, and seems to have employed some of Rousseau's ideas in the education of his son. As well as being tutored by his father, Robert Malthus was a pupil of the writer Richard Graves (1715–1804). Activities in an apparently happy youth included 'fighting for fighting's sake, without the least malice', drama, rowing, skating and cricket. He entered Jesus College, Cambridge in 1784. He studied mathematics, graduating as ninth wrangler in 1788 (i.e., coming ninth on the list of those with Honours in mathematics), but also studied classical and modern languages and history. Correspondence shows that he was anxious concerning a speech defect, and the effect that this might have on a career in the Church, but he seems to have read regularly in the college chapel, and to have won prizes for declamations in both Greek and Latin. On the other hand a visitor to his home in later life describes 'a defect in the palate' and mentions his speech as being 'hopelessly imperfect'.

He continued to read widely after graduating, including in the field of geography (letter dated 17 April 1788). He became a Fellow of his college in 1793. It is not absolutely clear when he entered holy orders (i.e., became a priest of the Church of England) – possibly it was late in 1788. But he became curate of Okewood, a chapel of ease in the parish of Wotton, Surrey, not far from his home, in 1793. These few years were spent in lively but good-humoured debates with his father, travel to Scandinavia and Russia (in 1799), and later (1802) to France and Switzerland, further study, and a little courting. In a letter dated 28 November 1800 he describes a visit to some 'pretty cousins' near Bath. He married one of them, Harriet Eckersall, on 13 March 1804: she was eleven years younger than he.

It was from this period that came some of Malthus' most important work. A pamphlet entitled *The Crisis* (apparently a plea for moderation at a time when oppressive measures were generating a real fear of revolution in England) was written in 1796, but represented a false start. He refrained from publishing it as the result of pressure from his father; some fragments eventually appeared elsewhere. The *Essay on Population* was first published in 1798, intended as something of a polemic against the utopian views set out by William Godwin's *Political Justice* (1793) and *Enquirer* (1797). His views were modified in the second edition of the *Essay*, published in 1803. Later editions contain modifications of less significance.

In the same year he was instituted as Rector of Walesby, Lincolnshire, a position he obtained through a family connection. He held the living (i.e., the rights and incomes from the parish) until his death, although he never resided there, paying out to a curate £70 (in 1826 increased to £80) of the £300 per annum he received from the office.

At the end of 1805 he became Professor of Political Economy and History at the newly founded East India College at Haileybury. This was one of the earliest appointments in political economy at a British institution. While at Haileybury he took services in the college chapel as well as lecturing. He claimed that, notwithstanding the alleged problem with his speech, he could make himself understood and, of his lectures, that his students 'did not even find [them] dull'!

Malthus held this post for nearly thirty years, publishing a series of articles and pamphlets on economic and political subjects, some of which were influential. The topics included the Corn Laws, the price of food, reform of the Poor Laws, population, and the nature of rent. None of these, however, had the long-term impact of the *Essay*. He did not travel extensively: he visited Ireland in 1817, and a further visit to the continent in 1825 was intended mainly to be for the benefit of his health, and that of his wife. He had a son and two daughters, one of whom died in childhood.

Although his work attracted astringent criticism, both in his lifetime and after, his views were in many ways quite moderate; 'Whiggish' politically, he adhered to the principles of the 'middle way' or 'golden mean'. He supported the emancipation of Catholics and popular education, deploring the extremes of *laissez-faire* capitalism, and the religious tests, for example for entry to the universities. His acquaintances found him extremely amiable. He was something of a moralist (although there is an eloquent description of the pleasures and beauty of sexual love – he uses the phrase 'virtuous love' – in the *Essay*). Patricia James (1979), on the basis of evidence in his travel diaries, describes him as being 'extremely susceptible to pretty girls'. Besides his father and his tutors, the writings of William Paley, Robert Wallace and Adam Smith were important influences on his thought and work.

Malthus was elected a Fellow of the Royal Society (FRS) in 1819, and became a member of the Political Economy Club, founded by Thomas Tooke, in 1821. He was elected one of the ten Royal Associates of the Royal Society of Literature in 1824, and, a few months before his death, was elected a Fellow of the Statistical Society. Towards the end of his life learned societies in France and Germany also honoured him. He advised Government, and gave evidence to a House of Commons Committee on Emigration in 1827.

He died, probably of a heart attack, on 29 December 1834, while spending Christmas at his wife's family home near Bath. He is buried in Bath Abbey, where a commemorative wall plaque describes him as:

> Long known to the lettered world by his admirable writings on the social branches of political economy, particularly the 'Essay on Population.'

> One of the best men and truest philosophers of any age or country ... His writings will be a lasting monument of the extent and correctness of his understanding.

2. Scientific Ideas and Geographical Thought

By far the most influential of the writings of Malthus was the *Essay on Population* (first edition 1798). At an early point in that essay (in fact a work of almost 400 pages) he succinctly sets out the notions for which he will ever be remembered, arguing from basic principles:

> I think I may fairly make two postulata. First, That food is necessary to the existence of man. Secondly, That passion between the sexes is necessary, and will remain nearly in its present state. (p. 11)

> Assuming then, my postulata as granted, I say, that the power of population is indefinitely greater than the power in the earth to produce subsistence for man. Population, when unchecked, increases in a geometrical ratio, Subsistence increases only in an arithmetical ratio. A slight acquaintance with numbers will shew the immensity of the first power in comparison with the second.
> By that law of our nature which makes food necessary to the life of man, the effects of these two unequal powers must be kept equal. This implies a strong and constantly operating check on population from the difficulty of subsistence. This difficulty must fall some where; and must necessarily be severely felt by a large portion of mankind. (pp. 13–14)

He later expands:

That population does invariably increase, where there are means of subsistence, the history of every people that have ever existed will abundantly prove.

And, that the superior power of population cannot be checked, without producing misery and vice, the ample portion of these too bitter ingredients in the cup of human life, and the continuance of the physical causes that seem to have produced them bear too convincing a testimony. (pp. 37–8)

'Famine and pestilence' he argues, therefore, are the mechanisms that provide the 'check' to the otherwise inexorable increase in numbers:

Famine seems to be the last, the most dreadful resource of nature. (p. 139)

The first edition of the essay seems to have been written at some speed, in answer to Godwin's treatise. He claimed that the only sources which he consulted were David Hume, Robert Wallace, Adam Smith and Richard Price. Although the argument is usually very clear, and the prose limpid, there is a good deal of repetition, and data are in some places lacking. Nevertheless he included some statistics on the numbers of births and burials over periods of some years for a number of states – Prussia, Lithuania and Pomerania, for example – attempting to demonstrate that during times of rapid population growth, certain years were 'very sickly'. On the other hand, he asserted that in the American colonies, and in the then recently independent United States, with an abundance of land, and rather fewer checks than in European countries, there was a population doubling time of about 25 years.

The pessimism implicit in the argument of the first edition, the apparent insuperability of the 'checks' to progress and improvement in the lot of humankind, attracted criticism. This hard line was tempered somewhat in the second edition, which appeared in June 1803. Continental travel and opportunities for further reading and discussion gave him access to much more information. In the new edition the 'checks' appeared not as insuperable obstacles to improvement and social progress, but as defining limits, as dangers that must be overcome if progress were to be achieved. Malthus always claimed that his was a practical, applicable approach. Effectively the second edition represented a different publication. Even the title was different. That of the first edition was *An essay on the principle of population as it affects the future improvement of society, with remarks on the speculations of Mr Godwin* ... By 1803 this had become *An essay on the principle of population, or a view of its past and present effects on human happiness, with an inquiry into our prospects respecting the removal or mitigation of the evils which it occasions*, indicating a rather more optimistic outlook. As one commentator (James Bonar) put it: '... the second essay ... lifts the cloud from the first.'

In the second edition of the *Essay* (and those that followed), Malthus emphasizes that the power of civilization is greater than that of population growth. He distinguishes between *positive* checks to population growth and *preventive* checks.

The positive checks to population ... include every cause, whether arising from vice or misery, which in any degree contributes to shorten the natural duration of human life. Under this head therefore may be enumerated all unwholesome occupations, severe labour and exposure to the seasons, extreme poverty, bad nursing of children, great towns, excesses of all kinds, the whole train of common diseases and epidemics, wars, plague and famine. (*Essay*, third edition, 1806, Vol. I, p. 19)

Under the heading of preventive checks he indicates 'vice', under which he seems to have included abortion, birth control, adultery and prostitution. He deplored all these, emphasizing *moral restraint* which he wished to '... be understood to mean a restraint from marriage, from prudential motives, with a conduct strictly moral during the period of this restraint'.

War, cannibalism and infanticide emerge as checks to population growth, and several chapters are devoted to a discussion of the various population checks operating in a number of lands, including Tierra del Fuego, New South Wales, New Zealand, North America, Siberia, the Pacific islands, China and Japan, as well as countries in Europe which he had by then experienced. Some of these accounts display an approach that would now be considered integrative and geographical, showing the relationships amongst environment (soils and climate for example), population and customs and activities. The second *Essay* was thus a much more mature work than the first, although perhaps lacking in the latter's freshness and directness. Alterations continued, partly in response to criticisms. For example, in the third edition 'an observation [is] added on the propriety of not underestimating the desirableness of marriage, while we are inculcating the duties of moral restraint.' He did not want to give the idea that he was opposed to marriage and family life *per se*. We may note that by the time the third edition was being prepared for the press he was himself in receipt of a good income and happily married to his beloved Harriet!

The first publication after the appearance of the first edition of the *Essay* was a tract entitled *An investigation into the cause of the high price of provisions*. This amounted to a criticism of the Poor Laws. He argued that the distribution of money or food to the poor by local communities from the proceeds of a Poor Rate simply caused the bidding up of the price of food and should be discouraged; hand-outs to the poor also encouraged them to breed. He did not lose the opportunity of arguing 'To what ... can we attribute the present inability of the country to support its inhabitants, but to an increase in population?' Needless to say, the views the pamphlet encompassed were not universally welcomed. To those who sided with the poor, Malthus was a Satan; to the middle classes suffering high taxes he was a saviour.

Other studies in economics and political economy followed, on rents, prices and on the Corn and Poor laws, culminating in his *Principles of political economy* in 1820. Some of these depended on his earlier theorizing on the subject of population. There are contradictions, not unexpected in a writing career of over two decades, and a measure of repetition. Sometimes a short publication reappeared as a chapter in a longer work. Many of his contributions had a significant impact on the discussions of economic principles at the time, and indeed on government policy. Although they contributed to the establishment of a theoretical basis for economic theory, none had the impact of, or aroused quite the controversy of, the *Essay*.

3. Influence and Spread of Ideas

There is no dispute that the influence of Malthus was profound: his name is perhaps better known than that of any other theorist of his generation. His work has affected the development of population geography, demography and the life sciences. It has influenced the direction of public policy in a number of countries. Dispute arises as to the extent to which he has been vindicated by subsequent events. In a passage that summarizes his views he pessimistically wrote:

> It is undoubted, a most disturbing reflection, that the great obstacle in the way to any extraordinary improvement in society is of a nature we can never hope to overcome. The perpetual tendency in the race of man to increase beyond the means of subsistence, is one of the general laws of animated nature, which we can have no reason to expect will change. (*Essay*, first edition, p. 360)

Passages such as the above are often quoted by the more extreme advocates of environmentalism and 'Green' political groups. The planet's population growth, they argue, must be controlled or the consequence will be unimaginable degradation. This type of thinking (or some variant of it) lies behind the ultimately abandoned Indian sterilization policy of the 1970s, or China's current (1990s) 'one child' policy.

Certainly populations around the world have continued to grow spectacularly, and in many cases more-or-less geometrically. The planet's population, in 1998 about 5700 million (5.7 billion), has doubled at least three times since the time of Malthus. In some countries the birth rate was still increasing in the early 1990s (e.g. Ivory Coast, Sierra Leone, Yemen, Mozambique, Mali, Angola, Laos). Locally, at least, Malthus' 'checks' of 'pestilence and famine' have exerted a terrible toll, where the events predicted or implied by Malthus have been replicated in reality. In Ireland, between 1846 and 1850, the combined effects of famine and the 'pestilence' of typhus following the spread of potato blight (*Phytophora infestans*) caused the deaths of 1.5 million people: a figure as high as 3 million deaths has sometimes been quoted. Many people emigrated. The population of the island fell from 8.175 million in 1841 to 6.624 million in 1851, the fall continuing to 5.5 million by 1866, and by the early twentieth century to 4 million. In 1998 it was at about the same level. India and China, or at least parts of them, have for long had high densities of population, and offer some spectacular figures for famine deaths. Some estimates have it that between February 1877 and September 1878 some 9.5 million persons perished from famine in North China. Some 'natural' disasters can at least partly be attributed to high population densities (e.g. flooding through the results of deforestation, and the need to cultivate intensively low-lying alluvial areas): 3.7 million died in the Hwang Ho floods of August 1931. Currently, some theorists point to the destructive power of AIDS, and the consequent reduction in the populations of appreciable areas of sub-Saharan Africa, as a vindication of Malthusian ideas. Certainly at some times and in some places the Malthusian purist has been able to provide examples where population growth has been checked by 'famine and pestilence', accompanying, as Malthus put it, 'vice and misery.'

Others take a different view, arguing that famine can often be shown to be at least partly due to human mismanagement, sometimes the result of authoritarianism. The Irish potato famine is seen by some as partly the result of landlordism and oppression. The famine in the USSR of 1920–1 in which some 5 million perished (along with subsequent periods of famine in the same country) is seen as being due to the incompetence and maladministration of the Communist government rather than high population densities. There is nothing 'natural' or inevitable about famine, the argument runs.

Undoubtedly, in many respects the implied predictions of Malthus have been wanting. To offer a single example, here he is on human life-span:

> With regard to the duration of human life, there does not appear to have existed from the earliest ages of the world to the present moment, the smallest

permanent symptom, or indication of increasing prolongation. (*Essay*, first edition, pp. 160–1)

In an era when infant mortality was still high (Malthus himself was to lose a daughter in childhood) and medical expertise, sanitation and public health were still crude, Malthus could not reasonably have predicted the developments in medical science and technology of the two subsequent centuries.

In that time many human societies have undergone a remarkable transformation – the *demographic transition*. As prosperity increases, the death rate falls; soon after, the birth rate drops, sometimes for several decades, until it reaches, or falls below, the death rate. Britain and the USA underwent this transition in the nineteenth century. The transition in 'developing' countries was still continuing in the 1990s. Even in Catholic countries such as Italy, Spain and Ireland, where opposition to contraception has traditionally been strong, the birth rate has fallen. With falling infant mortality, as parents see that there is every chance that all their offspring will reach healthy adulthood, they have fewer children. The desire for a consumerist, affluent lifestyle, the improvement in, and availability of, contraception, and improvements in education, especially that of women, may be other contributory factors. The apparent readjustment of rates of increase as populations pass through the transition has encouraged critics of Malthusian doctrines. Indeed, the falling of average numbers of births per woman below replacement levels and the 'ageing' of populations in many Western countries is of concern to a number of governments and UN agencies, and in some circles the notion of a 'population implosion' has replaced that of a 'population explosion'.

Malthus said rather little about the application of his ideas to natural populations of plants and animals, although in the first chapter of his 1798 *Essay* he states:

> Through the animal and plant kingdoms, nature has scattered the seeds of life abroad with the most profuse and liberal hand. She has been comparatively sparing in the food, and the nourishment necessary to rear them. The germs of existence contained in this spot of earth, with ample food, and ample room to expand in, would fill millions of worlds in the course of a few thousand years. Necessity, that imperious all pervading law, restrains them within the prescribed bounds. The race of plants and the race of animals shrink under this great restrictive law ... Among plants and animals its effects are waste of seed, sickness, and premature death. (pp. 14–15)

Yet it is in studies of animal populations that Malthusian mechanisms have been found to be most applicable. Both in experimental studies (such as those of G.F. Gause in the 1930s) and in field investigations it has been shown that organisms' populations sometimes increased rapidly in periods when resources allowed, only to 'crash' as the result of starvation and disease when numbers became too large. Thus in D.I. Rasmussen's 1941 classic description of 'Biotic communities of Kaibab Plateau, Arizona' (*Ecological Monographs*, No. 3, 229–75), a population of deer was revealed to have increased spectacularly over several decades in the absence of predators (which were artificially controlled) until, when food supplies became sparse, disease occurred and the population collapsed. (It should be noted that this study has been criticized more recently.) A Malthusian stance is implied by the title of David Lack's important and influential volume *The Natural Regulation of Animal Numbers* (Clarendon Press, Oxford, 1954), and indeed Lack specifically makes the link between the study of animal population dynamics and the prediction of possible human populations. Interestingly, quoting Malthus, the final paragraph of the book includes the statement '... temporarily, at least, our numbers seem to be reaching saturation'.

Another way in which Malthus influenced science was through the ideas of Charles Darwin. Malthusian doctrines were *de rigueur* in fashionable Whig salons in London in the 1830s, as Darwin returned from his *Beagle* voyage. Indeed, that radical, feminist propagandist for the Malthusian cause Harriet Martineau was, for a while, his elder brother Erasmus' girlfriend. There were a number of friends in common between the Darwin and Malthus families. Charles himself read the sixth edition (not the first, as is sometimes stated) of Malthus' *Essay* about September 1838, after his unpublicized 'conversion' to an evolutionary outlook eighteen months before, although he had never met Malthus. Malthusian logic runs powerfully through chapter 3 of *On the Origin of Species by means of Natural Selection* ... ('The struggle for existence'), which abounds with phrases such as 'geometrical ratio of increase' and 'checks to increase'.

Malthus emphasized that the human population was growing rapidly, and that humans were biological, as well as social, economical and political, beings, having a strong sex drive and requiring food. He has been one of the spurs of the environmental movement. But his critics have attacked him on a number of grounds. Karl Marx (1818–1883; this series, Vol. 19), in *Capital*, was a vigorous critic, arguing that poverty was the result of unjust social institutions and capitalism, and not growth in population. There were many who found Malthus' pessimism unappealing, particularly in the progressive, economically expanding, forward-looking nineteenth century. (To take a single example, the British statesman W.E. Gladstone read widely on the subject, and was a bitter opponent of Malthusianism.) Malthus was attacked for confusing moralistic and scientific approaches to population. He made much of attempting to control population growth, particularly amongst what he termed the lower classes and the poor, by the postponement of marriage – 'moral restraint'. Yet he opposed birth control vigorously, as he thought it would reduce the drive to work hard, particularly amongst the 'lower classes'.

Thomas Robert Malthus remained a devout priest of the Church of England until his death. Nevertheless his ideas were criticized as being anti-Christian or atheistical by some groups. Some looked to the Old Testament injunction to 'increase and multiply' and took exception to his notions of control on the rate of population growth. Like controversial writers before and since, he claimed in his lifetime that he was misquoted, and wrote to friends that he was attacked by those who had not read his work (e.g. in a letter dated 15 September 1834, three months before his death). But his material was written in such a manner as to attract criticism: it is but a short step from saying that smallpox is one of the checks to population growth to saying it is to be encouraged as such a mechanism; criticizing the mode of operation of the Poor Laws is easily confused with a suggestion that all provision for the poor should be abandoned. For the most part Malthus bore the attacks with dignity. The controversy contributed to his notoriety. When Harriet Martineau asked him if he ever 'suffered in spirits from abuse heaped on him', he replied 'Only just at first'. He had never 'after the first fortnight' been kept awake at night as the result of worry about what people felt about him.

The debate between the Malthusians and sceptics has waxed and waned since his day.

It was in American geography that anthropography and anthropogeography were taken up early in the twentieth century, as Davisian physiography lost ground. Some of these studies, for example those by Mark Jefferson, might be viewed as contributions to an embryonic population geography. Inevitably, in the absence of a readily available demographic literature, Malthus was further considered. In January 1925, at the Association of American Geographers meeting in Washington, DC, Mark Jefferson (1863–1949) stated: 'It is obvious that Malthus' reasonings were unsound.' He pointed out that few countries had regular census surveys in the 1790s, and Malthus did not have many reliable data for his assertions. Moreover, Jefferson carefully analysed census figures for many countries in Europe and the Americas (UK, France, Sweden, USA, Argentina), and showed that the doubling time had been progressively lengthening in the nineteenth and early twentieth centuries: there was no evidence of continued geometric rise. On the spectacular population decline of Ireland he pointed out that many 'Irish' lived elsewhere: 'There is no lack of Irish', he declaimed. Jefferson was also critical of Malthus' statement that food production must of necessity rise but slowly:

> Malthus did not know. He felt sure of many things. He thought and wrote much about the future but had no inkling of the new epoch that was about to dawn for men, the age of steam and steel, of engines and machines and motors, of capital organized, of tele-communication and transportation. And it was precisely these things, just on the modern side of his horizon, that have put food back into a minor place.

But such an interpretation was not universal. At the same meeting, the soils specialist C.F. Marbut (1863–1935) gave a paper entitled 'The rise, decline, and revival of Malthusianism in relation to geography and character of soils'. In this he pointed out that much of the great expansion in food production had come from the dark-soiled areas of the world, and argued that most of these were approaching their limits in production. He continued:

> If the law of diminishing returns be an inexorable one it seems only a matter of time when the law of material limitation of population will operate much more mercilessly than at any time in the past; for it is evident that the future contains lurking within it no possibility of such an increase in production as has taken place during the last half century through the utilization, for the first time in ... history [of] the black soils of the world.

Analogous examples from each side of the debate could be found from almost any decade over the last two centuries.

Sometimes the assertion made is that Malthus was 'too much a theorizer'. His work, particularly his early work, it is stated, was not founded on sufficient solid data. Comparison with Darwin or Marx shows him in an unfavourable light: there was no long period of careful accumulation and evaluation of information. Some authors have argued that Malthus' somewhat reactionary, moralistic views hampered the development of scientific demography rather than encouraging it. This is perhaps somewhat unfair: certainly his implied prophecies have not been fulfilled universally, and technology, medical science and social change have overtaken him. Nevertheless he established a definite conceptual framework for demographic studies, emphasizing processes and causes. Charles Darwin, whose natural selection owes so much to his reading of Malthus (consider the quotation on animal populations above), has been described as an integrator rather than an originator. So too Malthus erected a theoretical model combining the ideas of others. Sometimes it has been found useful: sometimes less so.

Bibliography and Sources

1. ARCHIVAL SOURCES

Although Malthus taught at the East India College at Haileybury for thirty years, Haileybury and Imperial Service College, as the institution has now become, currently has no archival material relating to him. The College does, however, have an excellent portrait, reproduced on page 57. Important documents are held in the Public Record Office and the India Office Records. Others are in the Guildford Muniment Room, the Berkshire Record Office and in private ownership. He was a prolific correspondent, and his letters are scattered. James (1979) gives very full details of archival sources.

2. SELECTED WORKS ON THOMAS ROBERT MALTHUS

Bonar, J., *Malthus and His Work*, George Allen and Unwin, London, 1885. (Second edition, 1926.)

James, P., *Population Malthus: His Life and Times*, Routledge and Kegan Paul, London and Boston, 1979.

Jefferson, M., 'Looking back at Malthus', *Geographical Review*, Vol. 15, No. 2 (1925), 177–89.

Keynes, J.M., *Essays in Biography*, Macmillan and Co., London, 1933, Part 2, chapter 1, 'Robert Malthus', 95–149.

Marbut, C.F., 'The rise, decline, and revival of Malthusianism in relation to geography and character of soils', *Annals of the Association of American Geographers*, Vol. 15, No. 1 (1925), 1–29.

Petersen, W., *Malthus*, Heinemann, London and Harvard University Press, Cambridge, Massachusetts, 1979.

Young, M., 'Malthus and the evolutionists: the common context of biological and social theory', *Past and Present*, Vol. 6 (1969), 304–47.

3. PRINCIPAL WORKS BY THOMAS ROBERT MALTHUS

1798 *An essay on the principle of population as it affects the future improvement of society, with remarks on the speculations of Mr Godwin ...* London. (Published anonymously.) (Reprinted with the original pagination by Macmillan, 1926 and 1966.)

1800 *An investigation into the cause of the high price of provisions*, London. (Published anonymously.)

1803 *An essay on the principle of population, or a view of its past and present effects on human happiness, with an inquiry into our prospects respecting the removal or mitigation of the evils which it occasions*, London.

1815 *An inquiry into the nature and progress of rent, principles by which it is regulated*, London.

1820 *Principles of political economy, considered with a view to their practical application*, London.

1823	*The measure of value, stated and illustrated, with an application of it to the alteration in the value of the English currency since 1790*, London.
1827	*Definitions in political economy*, London.

Chronology

1766	Born 17 February near Guildford, Surrey
1784	Entered Jesus College, Cambridge
1793	Elected Fellow of Jesus; appointed curate, Okewood, Surrey
1798	First edition of *Essay on population* published
1799	Travels in Scandinavia, Russia to collect data
1800	*Cause of the high price of provisions* published
1802	Travels in France, Switzerland to collect data
1803	Second edition of *Essay on population* published. Instituted as Rector of Walesby, Lincolnshire, 21 November; living held until death but never resided in parish.
1804	Married Harriet Eckersall of St Catherine's, Bath, 13 March
1805	Appointed Professor of History and Political Economy, East India College, Haileybury
1815	*Nature and progress of rent* published
1817	Visited Ireland
1823	Further travels on the continent, at least partly for the benefit of his health and that of his wife
1827	*Definitions in political economy* published
1834	Died 29 December near Bath; buried Bath Abbey, 6 January 1835

Patrick Armstrong teaches geography at the University of Western Australia. This chapter was written while on study leave from that institution, at St Deiniol's Library, Hawarden, Flintshire, North Wales.

Akira Nakanome

1874–1959

Hiroshi Ishida

Taken in 1918. Courtesy of Masakazu Morita

Akira Nakanome (1874–1959) was the second Japanese government scholar sent to Europe to study geography, and also the second founding professor of geography in higher education in Japan. The first professor was Naomasa Yamasaki (1870–1928), who organized the first geography programme in higher education, at Tokyo Imperial University. Akira Nakanome also co-operated with Takuji Ogawa (1870–1941) to found the first chair of geography at Kyoto Imperial University. Yamasaki is well known as the founding professor of scientific geography in Japan, but Akira Nakanome is now scarcely known among Japanese geographers. This chapter describes Nakanome's writings and evaluates his works mainly from a geographical viewpoint. Akira Nakanome was the predecessor of my own teacher, Professor Tokiwa Takao, and I was the fifth-generation Professor of Geography at Hiroshima Higher Normal College, and at the University of Hiroshima. When I was a student at the Higher Normal College of Hiroshima, I was told that the Professor of Geography named Nakanome was so talented in linguistics as to have deciphered peculiar letters discovered in a cave in Hokkaido. More than a quarter-century later, when I served in the Department of Geography, University of Hiroshima, no one passed on stories about him any more.

1. Education, Life and Work

Akira Nakanome was born into a middle-class Samurai family of the Sendai clan in 1874, soon after the Meiji Restoration (1868), when his father was appointed superintendent of Kurokawa county, Miyagi prefecture. Nakanome was taught Chinese classics in his childhood, and French early in middle school. All through his life he was distinguished by his language skills. After finishing the humanities course at the Second High School, with German as his major foreign language, he entered Tokyo Imperial University, again majoring in German. He had the exceptional honour of being granted a silver watch by the Emperor Meiji in 1899.

Nakanome assumed the professorship of German at the Fourth High School (old system), of which the President was Tokiyuki Hōjō (1858–1929). Before long President Hōjō was appointed founding President of the Higher Normal College of Hiroshima. He took several promising professors with him, among them Akira Nakanome, who was sent to study geography in Europe for three years.

Studying Geography Abroad

Akira Nakanome studied geography at the University of Vienna for his first two years under the tutelage of Albrecht Penck (1858–1945; see *Geographers*, Vol. 7), who was intensely occupied with completing his original study on glacial geomorphology. In the summer of Nakanome's first year at Vienna, a month-long geographical excursion was taken to the Austrian Alps, with Albrecht Penck as leader and Nakanome as his assistant. An American researcher, Ellsworth Huntington (1876–1947), of the Carnegie Institute, was also a member of the party. Carefully observing the glaciers and topography, the group strengthened their academic ties and friendship.

Penck's lectures on the regional geography of the Netherlands and Belgium were especially popular among the students, and Nakanome was quite attracted to Penck. Unfortunately for Nakanome, Penck was transferred to the University of Berlin, to succeed Ferdinand von Richthofen (1833–1905; see *Geographers*, Vol. 7). Penck's successor at Vienna was Eduard Brückner (1862–1927), a former student of Penck's. Thus Nakanome was guided during his third university year by Brückner, who greatly enthused him for glacial climatology. During his years at the University of Vienna, Nakanome studied geomorphology, geology, climatology, biology and ecology. He was also very interested in regional geography and glacial climatology. Geography alone did not claim all his interests, for he also read widely in the anthropological works of Friedrich Ratzel (1844–1904; see *Geographers*, Vol. 7) of the University of Munich, and visited him. Nakanome had missed a chance to meet Professor Richthofen, but he actively conferred with other distinguished professors, including Wilhelm Schmidt (1868–1954), the Austrian ethnographer and philologist. Nakanome paid special attention to Richthofen's work, because Penck had often referred to Professor Richthofen in his writings. Beyond his normal attendance at lectures and seminars, and library work at Vienna, Nakanome often went on geographical excursions and did fieldwork. When time permitted, he travelled widely throughout Europe, making the round trip by ship between Vienna and Budapest four times. During his first year he spent two months of summer vacation (9 July – 1 September) in the Alps. His second year saw a four-month trip to Carinthia in Austria, to Italy and adjacent areas, and a two-week trip down the Rhine. His third-year trips were yet farther away, to the Balkan states, Turkey and Russia for three months.

On these journeys Nakanome made keen observations of languages, ways of life and livelihoods in both urban and rural areas. He actively sought out many professors in countries other than Austro-Hungary and Germany: for instance, he attended lectures at the University of Rome. As is seen in his list of writings, he carefully kept a full diary of each journey; translation of these into English would be of great value to the history of geographic thought.

Geography Programmes in Japanese Higher Education

Akira Nakanome was one of the first two founding professors of geography programmes in Japanese higher education. As the first Professor of the geography programme within the Department of History and Geography (at one time,

Geography and Natural History) at the Higher Normal College of Hiroshima, he taught alongside other members of the faculty. Because of his distinguished achievements, his classes attracted many students, and his public lectures were popular with citizens and schoolteachers.

Geography at the Higher Normal College of Hiroshima was taught in close connection with history, civics and natural history. The geography programme itself comprised general geography, regional geography, and practice in techniques. Professor Nakanome was in charge of the programme as a whole, but preferred to teach students human and regional geography. Physical geography was taught mostly by Professor Manju Nishimura, who took up a lectureship at the Higher Normal College of Hiroshima in 1908, but was transferred to the Women's Higher Normal College at Tokyo in 1910. Physical geography included geomorphology, geology, climatology and astronomy, while human geography comprised historical geography, anthropogeography, political geography and economic geography. The regional geographies of various countries and continents, and of Japan, were also taught. Nakanome was interested as well in *Heimatkunde* (local geography, or homeland studies). Such geographical emphases were reflected in the educational offerings of middle schools and the normal schools.

2. Scientific Ideas and Geographical Thought

Nakanome's scholarly life may be examined in three aspects. First are his interests in climatic changes, anthropogeographical and linguistic studies and local geography. Second are his area-cultural and language studies. Finally, his biographical and local historical studies blossomed as he grew older.

His study of glaciology under Albrecht Penck in 1904 and 1905 and E. Brückner in 1906 generated a lifelong interest in the topic. But he was even more interested in glacial climatology, seen as the relation of glaciers to climatic change. Inspired by the Brückner cycle in particular, Nakanome tested ideas on climatic changes in Japan, and even applied Brückner's theory to explain the global movement of nomads. As time permitted, he actively conducted investigations into geomorphology and volcanism, studies that were full of his original views. However, Nakanome's academic skill was most fully displayed in anthropogeography. He led investigations not only into anthropological traits, but also into the languages of indigenous tribes such as the Tungus. Despite disadvantageous conditions just after the Russo-Japanese War, he made valuable observations and prepared accounts of the Tungus area. Further, his ability to learn languages was so great that he compiled a dictionary and grammar (in Japanese, and later in German) of the indigenous languages of Sakhalin. He delivered lectures on regional geography and directed the practice of geographical techniques during his tenure at the Higher Normal College of Hiroshima.

The significance of economic geography was being jointly advocated in Hiroshima by Professor Takujirō Hori, an economist, and Nakanome. It was Nakanome who initiated the fortnightly lectures in economic geography at Kyoto Imperial University which continued for several years.

Nakanome's lectures told of rare experiences and observations in several countries, and were filled with unique, original interpretations. His anthropogeographic view of the making of the Japanese nation seemed to have been too individual to be accepted by academic circles in the early twentieth century. To sum up this first phase of his activities, he was strict neither about the boundary between physical and human geography, nor about other boundaries within science, but was flexible in his thinking and behaviour.

Nakanome's talents were most effectively displayed as educator and scholar during his presidency of the Foreign Language College of Osaka, which he assumed in 1921. In Osaka Nakanome hosted many distinguished figures from other countries, among them Ellsworth Huntington in the summer of 1923. Huntington visited several large factories to observe the efficiency of workers. The following year Nakanome visited Yale University to renew his old friendship with Huntington.

Commanding several European languages – German, French, English, Spanish, Italian and Russian – and the neighbouring Chinese and Korean languages, he made as President various observational tours to the USA and Mexico, to East Africa, East Asia, and both South and South-East Asia. He was not satisfied merely to sit in the presidential chair, but travelled widely to observe topography, vegetation, landscapes, peoples and ways of life, cultural remains and education during his travels. He thereby realized a dream he had cherished since childhood. His tour reports varied in viewpoint and literary style from tour to tour, and any of these reports would have been deemed worthy as part of a doctoral thesis in regional geography by the international standards of that day.

Akira Nakanome was given the honour of presenting accounts of his study tour of East Africa at the general meeting of the Tokyo Geographical Society in 1928, an honour indicating the high evaluation of his achievements in geography and in education. He did not neglect a modest concern with geography, and published books on *Heimatkunde*, and on climate. His language abilities and interests in how to combine Latin letters in order to express Japanese pronunciation led him to contribute considerably to the controversy on language reform. His achievements and educational contributions to French and German were attested to by his receipt of an Award for National Cultural Merit from the French government in 1925, and the Order of Cultural Merit from the German government in 1933. From 1921 to 1933 his linguistic talents and world-wide activities were most evident. He had many essays printed at his own expense, to be distributed among relatives, friends and colleagues. Even after his retirement in 1937, he was not indifferent to the condition of geography in Japan. He advised geomorphologists no longer to adhere to the three volumes of *Das Antlitz der Erde* by Eduard Suess (1831–1914), or to Albrecht Penck's *Morphologie der Erdoberfläche* of 1894.

With ageing, Nakanome's identity with his home province became stronger, and he turned to the revival of clan identity and a strong interest in his home province's local history. He was appointed chief editor of *A History of Miyagi Prefecture*, which saw publication after his death. His study of Rokuemon Tsunenaga Hasekura (1571–1622), an ambassador sent to the Pope by Daimyo Masamune Date, saw publication in 1957. Nakanome's interest in Hasekura was so strong that he made journeys to collect sources relating to Hasekura in Italy, Spain, Mexico and the Philippines – all places where Hasekura had travelled.

At the age of 82, in 1957, Nakanome took university students to observe the location of landslides which had taken place in the valley of the Arao river, along the road to the Onikubi dam. He discovered that these phenomena were connected with glacial deposits, and that the glaciation in the coldest period extended as far down as Iwade town. It is thus seen that he did not support the theory of 'glaciation in the highest altitudes only'.

Rather than being primarily interested in geomorphological evidence of glaciation, he was more interested in climatic change as shown by the advance and retreat of glaciers. He discussed famines and the movements of people on a global scale, as related to climatic changes. He put forth an interesting discussion of historical and cultural traits supporting the cultural diffusion theories of Eduard Brückner and Friedrich Ratzel. Nakanome was the first scholar to test Brückner's cycle in East Asia and Japan.

3. Influence and Spread of Ideas

Akira Nakanome was more of a fieldworker and traveller than a geographical methodologist. He travelled on horseback in Africa, hired a horse-wagon in the Balkans, and moved by canoe in northern Sakhalin. Nakanome's studies were cross-cultural and interdisciplinary efforts in which he utilized his command of several modern European and Asian languages, as well as indigenous tongues. Several students, including Takeshiro Nakano, became geography professors, but Nakanome's influence was not confined to geographical circles but was even stronger in educational and linguistic circles. With regard to his geographical outlook, scholars now regard his anthropogeographical approach highly, particularly his ideas on how to seize upon the character of a region, and how to describe it. His regional works are worthy of reconsideration. It is generally known that Michitoshi Odauchi and others promoted the homeland studies movement in Japan. But it is scarcely noted today that Nakanome inspired Odauchi when he was a student in Hiroshima. Certainly culture-area studies in Japan have their roots in Nakanome's painstaking and time-consuming work. Finally it should be noted that Nakanome was a pioneer of cross-disciplinary studies in Japan.

Nakanome published more than ten books, most of which are difficult to find today. Many of his works remain unfinished. It is fortunate, therefore, that the entire corpus of one of his books has very recently been included in *Ethnology of Northern Tribes: The Tribes of Sakhalin and the Kuril Islands*, edited by Ken'ichi Tanigawa.

Many works of reference list Nakanome as an educator and linguist. Yet an obituary in the *Kahoku Shinpo Press* recognized him as a geographer and linguist who studied glaciers abroad for three years. Even after all his papers and his library had been donated to the Juan Library (named in memory of the martyr Juan Gotó, sixteenth to seventeenth centuries), Albrecht Penck's portrait, along with Penck's letter to him dated 19 June 1909, remained pinned up on the wall of his study.

Acknowledgements

The author acknowledges the assistance of a number of persons and institutions. These include: Mr Iyohiko Nakanome, Professor Shimon Meguro, Professor Forrest Pitts, the University of Hiroshima Research Centre for Regional Geography, the Foreign Language University of Osaka, the Municipal Museum of Sendai, the Library of the University of Fukuyama. The photograph of Nakanome (which was taken in 1918) was provided by Masakazu Morita.

Bibliography and Sources

1. OBITUARIES AND REFERENCES ON THE LIFE OF AKIRA NAKANOME

Kahoku Shimpo Press, 28 March 1959:

Yamada, T., 'Teacher Akira Nakanome: the man';

Ohara, H., 'Recollections of teacher Akira Nakanome';

Yamasaki, Y., 'Mr Akira Nakanome, the author of *The Alps and the Rhine River*';

Nasukawa, H., 'A sketch of teacher Akira Nakanome'.

Nakanome, A., *Recollections of My Life over Eighty Years*, 1957. Publication commemorating the 35th anniversary of the founding of the Foreign Language University of Osaka.

2. THEMATIC BIBLIOGRAPHY OF WORKS BY AKIRA NAKANOME

Anthropogeography

1914	'Karafuto no shaman-kyō' (Shamanism in Sakhalin), *Gakkō kyōiku (School Education)*, Vol. 1, No. 1.
1914	'Karafuto dojin no hanashi' (Facts concerning Sakhalin aborigines), *Shigaku kenkyū (Historical Review)*, Fusanbo.
1914	'Karafuto dojin no gengo' (Languages of Sakhalin aborigines), *Geimon (Arts)*, Vol. 5, 11.
1916	'Nikubun zoku no meisho' (The name of the Nikvn tribe), *Geimon*, Vol. 7, 10.
1916	'Karafuto no dojin' (Aborigines in Sakhalin), *Gakkō kyōiku*, Vol. 3, 12.
1917	*Nikubun bunten (Grammar of Nikubun)*, Sanseido.
1917	*Orokko bunten (Grammar of Orok)*, Sanseido.
1917	*Karafuto no hanashi (Facts on Sakhalin)*, Sanseido.
1917	'Tsungusu-go no tango hikaku' (A comparative study of Tungus words), *Geimon*, Vol. 8, No. 2.
1917	'Waga kuni ni hozon sareta kodai Toruko moji' (Ancient Turkish letters preserved in Japan), *Shōko (Classicism)*, Vol. 71.
1918	*Dojin kyōiku ron (On the Education of Aborigines)*, Iwanami Shoten.
1918	'Hokkaidō temiya dokutsu Makatsu-go boshi ni tsuite' (An epitaph in the Makhats language discovered in Temiya Cave in Hokkaido, Nos 1, 2), *Rekishi to chiri (History and Geography)*, Vol. 1, Nos 1, 2.

Climatic Change

1917	'Hyōga to kikin' (Glaciers and famine), *Geimon*, Vol. 68.
1919	'Chū-a no kikō hendo to waga kuni e no eikyō' (Climatic changes in central Asia and their impact on Japan), *Shōko*, Vol. 75.
1932	*Kikō to rekishi (Climate and History)*, Osaka Asiatic Society.

Historical Geography

1912   'Chizu ni araware taru Hokkaidō oyobi Karafuto' (Hokkaido and Sakhalin as they have appeared on maps), *Shōko*, Vol. 50.

1915   'An account of a journey to southern Kyushu', *Shōko*, Vol. 60.

1915   'Kagoshima, Miyazaki ryokō dan' (An account of a journey to Kagoshima and Miyazaki), *Shōko*, Vol. 60.

1918   'Karafuto no shokumin' (Colonization in Sakhalin).

1933   *Osaka kyōdo chiri jō* (An Introduction to the Homeland Geography of Osaka), Hakata Seisho-do.

1933   'Rekishi-jo no Nichi Hi kankei' (Japanese relationships with the Philippines in the historical past), *Nanyō kenkyū (Studies in the South Seas)*, Vol. 6, Foreign Language College of Osaka.

1933   'Ruson ni okeru Tsunenaga Hasekawa' (The facts concerning Tsunenaga Hasekura), *Supeingo bu zasshi (Spanish Bulletin)*, Vol. 5, Foreign Language College of Osaka.

1934   *Ruson Kikō (An Account of a Journey to the Philippines)*, Osaka Geographical Society.

1957   *Rokuemon Hasekura*, Hakurei-kai.

1960   (chief ed.) *A History of Miyagi Prefecture*.

Regional Geography

1910   'Tō-a ryokō dan' (An account of a journey in East Asia), *Chiri rekishi gakkai-shi (Bulletin of the Society for Geography and History)*, Higher Normal College of Hiroshima.

1916   'Tōbu Shiberia no senjō ni tsuite' (Battlefields in eastern Siberia), *Shokō*, Vol. 74.

1925   'Bei-Boku kenbun-roku' (A record of personal experiences in the USA and Mexico), *Kaigai shisatsu (Observation Abroad)*, Vol. 5, Foreign Language College of Osaka.

1928   'Afuriku shisatsu dan' (Report of an observational tour in Africa), *Chigaku zasshi (Journal of Geography)*, 473–5.

1929   'Chōkō yu-ki' (Report on a study tour of the Yangtze Kiang River), *Kaigai shisatsu*, Foreign Language College of Osaka.

1930   'Hōjin sekai isshū no tancho' (The first around-the-world trip by a Japanese), *Chigaku zasshi*, 500.

1931   *Barukan ryokō (An Account of a Journey Through the Balkan States)*, Osaka Geographical Society.

1932   'Itariya nikki' (Diary of a Journey in Italy), *Kaigai Shisatsu*, Foreign Language College of Osaka.

Miscellaneous

1919 'Furansu ni okeru chūtō gakkō chiri jigyō yōmoku' (Geographical syllabuses in the middle and high schools of France), *Rekishi to chiri*, Vol. 3, No. 1.

1921 *Chiriteki shigeki (Geographical Impulses)*, privately printed.

1925 *Seijin kyōiku to kōminka (Continuing Education and Civics)*, Foreign Language College of Osaka.

1931 *Nakanome godai no katei kyōiku (Home Education of Five Generations of the Nakanome Family)*, privately printed.

1932 *Shin gairai-go jiten (A New Dictionary of Words of Foreign Origin)*, Hakata Seisho-do.

1932–6 *Hakurei jihō (Personal Essays)*, Nos. 1–10, Hakurei kai.

1937 'Chiri kyōiku ni tsuite' (On geographical education), *Chireki kōmin gakkai-shi (Bulletin of the Society for Geography, History and Civics)*, Vol. 5, Higher Normal College of Hiroshima.

Chronology

1874 Born May 23 in Sendai

1896 Completed high school

1899 Graduated from Tokyo Imperial University. Appointed Professor of German at the Fourth High School

1903 Appointed first Professor of Geography at the Higher Normal College of Hiroshima. Received a government scholarship to study geography abroad for three years. Travelled to Austro-Hungary to attend the University of Vienna

1904 Geographical excursions in the eastern Alps and Switzerland

1905 Travels in Yugoslavia and Italy, and down the Rhine

1906 Travels through the Balkan states, Russia and Turkey

1907 Returned to Japan

1908 The earliest chair of geography in Japan founded at Kyoto Imperial University

1909 Journey to Korea. Akira Nakanome and Naomasa Yamasaki hosted Albrecht Penck in Kyoto and Tokyo

1910 Part-time lecturer at Kyoto Imperial University

1912–13 Fieldwork in Sakhalin

1916 Journey through Korea, China, and the Vladivostok area

1917 Publication of *Nikubun bunten*, *Orokko bunten* and *Karafuto no hanashi*

1918 Publication of *Dojin kyōiku ron* and *To-Ō nikki*

1919	Founding Vice-President of Matsuyama High School
1920	Publication of *Alps-zan to Rhine gawa*
1921	Founding President of the Foreign Language College of Osaka. Publication of *Chiriteki shigeki*
1923	Hosted Ellsworth Huntington in Osaka
1924	Journeys in the USA and Mexico
1925	Granted the Award for Educational Services by the Government of France. Publication of *Seijin kyōiku to kōminka*
1927	Journeys in East Africa. Founding of the Literature and Science University of Hiroshima
1928	Journey to China
1931	Publication of *Barukan ryokō*
1932	Granted the Award for Cultural Merit by the government of Germany. Publication of *Kikō to rekishi*
1933	Journey to the Philippines. Retirement from the Foreign Language College of Osaka. Publication of *Osaka kyōdo chiri jō*
1934	Publication of *Ruson kikō*
1939	President of the Academy for the Development of Asia
1957	Publication of *Rokuemon Hasekura*
1959	Died 27 March in Sendai city
1960	Publication of *A History of Miyagi Prefecture*

Hiroshi Ishida is Professor Emeritus of Geography at the University of Hiroshima.

John Ogilby

1600–1676

Charles W. J. Withers

Portrait by Peter Lely, engraved by Peter Lombart

John Ogilby was a translator, dancer, theatre manager, compiler and publisher of atlases, a map-maker and the author of *Britannia* (1675), a text generally recognized as one of the great works of chorography and cartography in Restoration science. Ogilby's importance rests less in his original contribution to geography – many of his great atlases, for example, were drawn from the work of others – and much more from his position as a publisher-courtier and as part of those crucial networks of royal patronage through which early modern science, including geography, was undertaken. He bore, at various times, the titles Master of His Majesty's Revels in the Kingdom of Ireland, Master of the King's Imprimeries (the King's Printer), and His Majesty's Cosmographer and Geographic Printer. He was a key figure in that circle of *virtuosi* helping plan and map London after the Great Fire of 1666. In his scheme to undertake a survey of England and Wales of which his 1675 work is the only published outcome, Ogilby's use of circulated queries and his employment of men such as John Aubrey and Gregory King shows him to be demonstrating that interest in national knowledge through empirical and political survey that characterized much late-seventeenth-century geographical enquiry.

1. Education, Life and Work

John Ogilby was born near Dundee on 17 November 1600. Nothing is known either of his parents or of Ogilby's early education, and we know the date of his birth only from later statements by his contemporaries John Aubrey and Elias Ashmole. Aubrey tells us only that Ogilby was from 'a gentleman's family and bred to his grammar', a remark which is suggestive, nevertheless, of Ogilby's social status. It is likely that the family moved to London soon after the Union of Crowns in 1603, perhaps as part of the retinue of followers of James VI and I. Ogilby's father was admitted to the Merchant Taylors Company in London in 1606 and we know that John Ogilby was likewise admitted, on 6 July 1629. Aubrey tells us that the young Ogilby paid his father's debts with money secured from a lottery managed by the Virginia Company in March 1612: lotteries were later to be an important source of funding for Ogilby's own publishing schemes. We know, too, that as a young man Ogilby was apprenticed to John Draper, a London dancing master and that, from the early 1620s, Ogilby was involved in several Jacobean courtly masques and dances. He is thought to have danced, for example, in Ben Jonson's 1621 masque *The Gypsies Metamorphosed*, and it may have been there that Ogilby injured himself and was lame thereafter. Certainly, the end of a potential career in choreography was to the longer-term benefit of his interests in chorography.

In general terms, it is possible to see several phases to Ogilby's varied career: dancing master, courtier and theatre owner between about 1620 and 1641; poet and translator from 1649; and, from about 1669, compiler of geographical works and atlases, culminating in his *Britannia* (1675). Throughout, he retained and developed the support of several patrons, not least that of the King himself.

The details of Ogilby's life in the later 1620s are uncertain, although it is possible he soldiered. We do know he moved to Ireland in the summer of 1633, under the patronage of Thomas, Viscount Wentworth, later Earl of Stafford, to become Master of the King's Revels in Ireland. In that capacity, Ogilby was responsible for the opening of Ireland's first theatre, the New Theatre in Dublin's Werburgh Street, sometime in 1637. This closed with the Irish rebellion in 1641. A short period of service to the Earl of Ormond ended with Ogilby's return to England about 1644, first to Cambridge and then to London. He married Christian (Katherine) Hunsdon on 14 March 1650 in St Peter-le-Poor parish in London. She was the widow of one Thomas Hunsdon, a Merchant Taylor, of Blackfriars in London. William Morgan, the son of her married daughter Elizabeth, was later to be involved with Ogilby in mapping and publishing projects and was Ogilby's executor after his death.

In 1649, Ogilby published his first translation, of Virgil. This was followed in 1651 by Aesop's *Fables* and by further translations of Virgil in 1654 and 1658. Ogilby's translations, particularly the 1654 folio, were splendid and carefully executed productions. His style was direct and true to the original and he paid great attention to paper quality and to the visual elements of printing. Later translations – of Homer's Iliad (1660), of Aesop's *Fables* and his *Aesopics* (1665 and 1668, respectively), of Homer's *Odyssey* (1665), and of Virgil in 1666 – are magnificent examples of seventeenth-century printing. Ogilby was also involved in the production of a two-volume edition of the Bible which he illustrated with a variety of chorographical scenes, and which was, like his Iliad, dedicated to Charles II. Ogilby's translations of the classics should not be seen as just an individual, or, even, as a particularly British, enterprise. He drew upon the translations of Virgil by Spanish scholars like the Jesuit de la Cerda and upon others such as Nicholas Caussin. In such ways, classical learning, book production and courtly patronage not only promoted elite authority but helped establish a literate public sphere across late seventeenth-century Europe.

Ogilby's support from and for royalty is most clearly evident in his 1661 *The Entertainment of his Most Excellent Majestie Charles II*. This was, essentially, a commemoration of the new King's coronation. In drawing, however, upon earlier traditions of triumphal processions, Ogilby used the streets of London as a theatrical stage for the promotion, through geography and spectacle, of ideas of British empire, civic virtue and royal authority. Ogilby successfully petitioned the King in order to secure a monopoly on his *Entertainment*, and probably secured the title of Master of the King's Imprimeries in the same year. In the summer of 1662 he returned to Ireland, again to found a theatre. By late 1665 or early 1666, however, Ogilby was again settled in London. He was appointed one of the city's assistant surveyors after the Great Fire of London (in which Ogilby is said to have lost £3000 worth of his own stock). This position brought him into touch with, amongst others, Robert Hooke, Christopher Wren, John Aubrey and Gregory King.

Ogilby's series of atlases and geographical compilations effectively began in 1669 with his *Embassy to China*. Certainly, it is from this period that he conceived of a series of atlases to cover the whole world, to be funded through lotteries, subscription plans and advertisements. The first, *Africa*, appeared in 1670. Others followed soon after: *Atlas Japannensis* (1670), *America* (1671), and *Atlas Chinensis* (1671), and *Asia* (1673). These works were not the fruits of his own labours so much as carefully edited and well-produced compilations of others' accounts, often in the face of competition from other authors such as Richard Blome, whose own *Britannia, or a Geographical Description of England, Scotland and Ireland* was published in 1673.

Ogilby's *Britannia* (1675) was much more securely based on contemporary and collaborative research. He again drew upon the support of the King and other noble patrons in the publication of the work, and upon the capacities of many of his scientific contemporaries for the survey work on which it was founded. The work, chiefly a road atlas of England and Wales with additional comments on the agricultural condition of adjoining land, marks the first major advance in British cartography since the Tudor period, even although it clearly drew upon earlier chorographical traditions. Ogilby's work was continued after his death, on 4 September 1676, by William Morgan. Ogilby was buried in the vaults of St Bride's Church, London.

2. Scientific Ideas and Geographical Thought

Ogilby's significance in this respect rests in four areas: his 1661 *Entertainment*; his production from 1670 of the several great atlases; his involvement in the survey, planning and mapping of London after 1666; and his 1675 *Britannia*. Only in the first and most clearly in the last of these, however, can we see Ogilby's own thoughts and ideas at work. What connected them all was his involvement in courtly patronage networks and in polite society and the emerging public sphere of the late seventeenth century. This background is important not just to any appreciation of Ogilby's works but in understanding the social context within which the practices of geographical enquiry then took place.

The full title of Ogilby's 1661 work to celebrate the restoration of the monarchy was *The Relation of His Majestie's entertainment passing through the City of London, to his coronation: with a description of the triumphal arches, and solemnity*. A second edition was produced in 1662. In this work, Ogilby drew upon accounts of the triumphal entry into London of James VI and I in 1604, the year after the Union of Crowns of England and Scotland. King James's procession was about establishing the British Empire from within, by symbolically representing the United Kingdom. Ogilby's own work likewise stressed the theme of Britain's empire, but on a grander scale, and he saw Charles II's authority as essential to both domestic loyalty and Britain's overseas expansion. These issues are clear from Ogilby's description of two of the four arches built for the 1661 procession:

> On the South Pedestal [of the first arch] was a Representation of *Britain's* Monarchy, supported by *Loyalty*, both Women. *Monarchy* in a large purple

Robe, adorned with Diadems and Scepters, over which she had a loose mantle edged with blue and Silver Fringe, resembling Water, on her Mantle the map of Great Britain, on her Head *London*, in her right hand *Edinburgh*, in her left *Dublin*.

In the second arch, pedestals in the upper stories 'were adorn'd with living Figures, representing *Europe*, *Asia*, *Africa*, and *America*, with Escutcheons and Pendents, bearing the Arms of the Companies Trading Into those parts'. In four further niches within this arch stood four women 'representing Arithmetick, Geometry, Astronomy, and Navigation'.

In such ways, Ogilby's 1661 work might be said to have *performed* British ideas of *imperium* and of the world as Britain's emporium, and, further, to have done so for those rationalizing practices like geometry and navigation then central to the advance of the new science, overseas travel, and the work of the Royal Society. In this case, he did not undertake such work himself: rather, he was, as was commonplace in this period, symbolically staging and embodying such matters for popular audiences in order both to impart a political message and to secure personal status.

Ogilby lost most of his stock in the Fire of London (in the Preface to his 1671 *Africa*, he tells us he was only worth £5 after the fire), and it may have been financial pressures that directed him towards his atlas projects. In 1669, Ogilby published *An Embassy from the East India Company of the United Provinces to the Grand Tartar Cham Emperour of China* (the work is known usually as *Embassy to China*). The work was largely derivative, principally from the work of the Dutch scholar Jan Nieuhoff, although Ogilby drew also from Athanasius Kircher and others. For several reasons, this work is important to any understanding of Ogilby's ideas and geographical thought. First, in being derivative of others, it was not untypical of the time. The fact that Ogilby and others drew from others' work should serve to remind us not just of the growing popular and public interest in geographical matters in this period, but of the competitive personal rivalries and intellectual markets that underlay the production of geography. Second, public acclaim for this work allowed Ogilby to adopt the same methods for the later atlas projects. Third, the work was largely funded by lottery. If not an original thinker in the usual understanding of that term, then, Ogilby was certainly astute as a geographical publisher in creating an audience for his books.

Africa, printed in 1670, was the first volume of what Ogilby intended to be a great Atlantic or English Atlas, a project shared (and likewise never realized) by others at this time such as Moses Pitt. It is useful, too, in containing a partial autobiography of Ogilby. Like the following atlases, the work was lavishly illustrated with engravings and maps. The atlases combined the latest travel accounts with Ogilby's own sense of the utility of survey as a means to national knowledge, for example, and comments on earth history and natural occurrences like earthquakes. In the prefatory pages to his *America*, Ogilby acknowledged his use of over 150 authors: many were classical authorities, others contemporary writers. In such ways, Ogilby, in the atlases particularly, should be seen as both testing the claims of ancient authors in relation to that new evidence about the geography of the world that, ship by ship, was arriving almost daily, and providing an accessible digest of the latest geographical facts and travels for popular consumption.

Ogilby's contribution to survey and mapping work as an assistant surveyor in London after 1666, and, later, his own *Britannia*, took place within a society increasingly aware of the contribution of the surveyor to economic progress. This was apparent in London, of course, but was evident also in the increase in number and competence of surveying manuals and in the social status of the surveyor. *Britannia* emerged from what was intended to be the fifth volume of his atlas series. A royal warrant shows that by late August 1671 Ogilby had secured royal approval for a survey of Great Britain. An advertisement of November 1671, and a prospectus entitled 'Mr. Ogilby's Design for Carrying on His Britannia', most likely dating from early 1672, show Ogilby undertaking his work through an appeal to appropriate persons to provide him with information. Ogilby invited 'all persons of Quality whatsoever ... to communicate to him by writing or otherwise, ... all such Remarques in their respective Counties as may be conducible to the said work. And for the further illustration thereof, those of the Nobility and Gentry who are desirous to have an account of their Seats and Families, with their Atchievements, inserted, may be pleased to repair, or send their Directions, to the forementioned places.'

In writing thus, Ogilby was clearly influenced by earlier traditions of regional writing, notably by men such as John Norden, whose own *Speculum Britanniae*, an attempted historical and geographical description of Britain, appeared only in two volumes, for Middlesex (1593) and Hertfordshire (1598). Yet his work is also to be understood as contemporary chorography through local knowledge. This is geography as an empirical enquiry from persons whose social status and noble birth suggested them to be credible sources. It was also work informed by actual survey, which began with the counties of Kent, Middlesex and Essex, most probably in 1672, and which certainly involved Gregory King, who was an important link between Ogilby and the Royal Society. Circulated queries, of which two drafts dating from 1672 and 1673 survive, show Ogilby and men including Christopher Wren, Robert Hooke and John Aubrey, as well as King, utilizing for *Britannia* that form of national knowledge through questionnaire that was used by Sibbald and Adair (this volume) in Scotland, by Robert Plot for Oxfordshire and Sir William Petty in Ireland.

The project as a whole remained under-funded, however, and the sheer scale of the task at the detail intended probably demanded that pragmatic decisions were reached to reduce the undertaking. *Britannia* is important, however, as a local document, because it provides systematically-collected information, and because in being undertaken by qualified practitioners using a simple set of repetitive techniques and standard queries, we can see in it geography's place in the emergence of those empirical practices then characterizing 'modern' science. The frontispiece illustration, by the engraver Wenceslaus Hollar, symbolizes splendidly the complex production of geography in these ways: a triumphal arch has the royal flag atop it; gentlemen riders set forth with maps in hand; others, seated at a table on which lie compasses and other instruments, study the globe; yet others are at work measuring the fields. It is not fanciful, indeed, to see the illustration as both chorography and choreography: expressive of Ogilby's interest in taking the appropriate steps to know one's national space.

3. Influence and Spread of Ideas

Ogilby's influence rests chiefly with his *Britannia* and in the cartographic 'tradition' the work prompted. Yet it is also possible to see an influence both in terms of his creating and reflecting public interest in geography, and in his helping place geographical enquiry through local survey within the development of contemporary scientific methodology.

Harley has suggested that Ogilby's *Britannia* had lasting influence upon later cartographic practice for four reasons. First, in order to meet the demand for a usable version of the larger work, Ogilby, in association with Morgan and Hooke, produced a version of the work without the road maps: this, published in 1676, was known first as *Mr. Ogilby's Tables of his Measur'd Roads*, and later as *Mr. Ogilby's and William Morgan's Pocket Book of the Roads*. Over twenty editions were produced and the work was still in print a century later. Second, there were the pocket-form reductions of Ogilby's texts and maps, prepared by other cartographers, based either on the texts, or on the maps alone. The maps-based volumes were more prolific and longer-lasting: the most popular was *Britannia Depicta or Ogilby Improv'd*, engraved by Emanuel Bowen, which first appeared in 1720 and went through thirteen subsequent editions. Third, his maps provided a basis for later Georgian map-makers. Fourth, his style of ribbon maps was popular elsewhere within Europe and, indeed, in North America.

In being involved with the survey of London as well as with *Britannia*, Ogilby was also influential in establishing the credibility and respectability of map-making in Restoration England, particularly within the royal court and within the emerging public sphere of that period. His several atlases should also be seen as items with contemporary cultural value, a sort of display value beyond the geographical information they contained. This is clear not just from the patronage networks that helped fund the books, and from the quality of the production, but from the prices paid for copies of his work at later seventeenth-century auctions. His importance rests, then, in his helping create, in the several ways documented here, an audience for geography.

Bibliography and Sources

1. SELECTED BIBLIOGRAPHY OF WORKS ABOUT JOHN OGILBY

Aubrey, J., *Letters by Eminent Men and Lives of Eminent Men*, John Murray, London, 1813, Vol. III, 466–70.

Fordham, Sir H.G., 'John Ogilby (1600-1676)', *Transactions of the Bibliographical Society*, N.S., Vol. VI (1925), 157–78.

Harley, J.B., 'Introduction', *Britannia*, Facsimile edition, 5th Series, Vol. II, Theatrum Orbis Terrarum, Amsterdam, 1970.

Taylor, E.G.R., 'Robert Hooke and the cartographical projects of the late seventeenth century', *Geographical Journal*, Vol. 90 (1937), 529–40.

Van Eerde, K.S., *John Ogilby and the Taste of His Times*, Dawson, Folkestone, 1976.

There is no single collection of Ogilby's manuscript material, although there are holdings with the Bodleian Library, the British Museum and Yale University Library. For a discussion of the extant manuscript material, see the bibliographical essay in Van Eerde (1976), pp. 173–9, particularly pp. 173–5. There are a number of portraits of Ogilby, the earliest dating from 1649. Both that image drawn by Peter Lely and engraved by Peter Lombart (see p. 77), and a later image drawn by Lely and engraved by William Faithorne in 1660, include the arms of Scotland: it is not clear that Ogilby intended to establish his identity as a Scot in doing so.

2. SELECTED BIBLIOGRAPHY OF WORKS BY JOHN OGILBY

1661   *The Relation of His Majestie's entertainment passing through the City of London, to his coronation; with a description of the triumphal arches, and solemnity*, T. Roycroft for R. Marriott, London.

1669   *An Embassy from the East India Company of the United Provinces to the Grand Tartar Cham, Emperour of China*, John Ogilby, London.

1670   *Atlas Japannensis: being remarkable Addresses, by way of embassy, from the East India Company of the United Provinces to the Emperor of Japan, English'd and Adorn'd by J. Ogilby*, John Ogilby, London.

1670   *Africa*, John Ogilby, London.

1675   *Britannia, Volume the First: or, An Illustration of the Kingdom of England and Dominion of Wales: by a Geographical and Historical Description of the Principal Roads thereof*, John Ogilby, London.

1676   *Mr. Ogilby's Tables of his Measur'd Roads*, William Morgan, London.

Chronology

1600   Born on 17 November, in the village or settlement of 'Killemeane' (now not traceable), near Dundee

c. 1603–4   Moved to London

1621   Danced in Jonson's *The Gypsies Metamorphosed*

1629   Admitted as member of the Merchant Taylors Company, London, 6 July

1633   Moved to Ireland, under the patronage of Thomas, Viscount Wentworth, as Master of the King's Revels

1637   Established and opened Ireland's first theatre

1644   Returned to England (known to be resident first in Cambridge and then in London)

1649   Publication of his first translation, of Virgil

1650   Married Christian (Katherine) Hunsdon, 14 March, in St Peter-le-Poor parish, London

1661   Publication of *The Entertainment of his Most Excellent Majestie Charles II*. Formally re-confirmed as Master of the Revels in Ireland; probably appointed that year to the position of Master of the King's Imprimeries (King's Printer)

1662   Moved to Dublin

1665–6   Returned to London

1669   Publication of *Embassy to China*

1670   Publication of *Africa* and *Atlas Japannensis*

1671 Publication of *America* and *Atlas Chinensis*, and began thinking about the project that resulted in *Britannia*; secured the title His Majesty's Cosmographer

1673 Publication of *Asia*

1675 Publication of *Britannia*

1676 Died 4 September 1676

Charles W.J. Withers is Professor of Geography at the University of Edinburgh.

Thomas Pennant

1726–1798

By Thomas Gainsborough. Courtesy of the National Library of Wales

Colin Thomas

Eighteenth-century Europe retained profound social and economic inequalities inherited and entrenched from medieval times, while simultaneously being subject to new forces of disruption as mercantile capitalism and urban industrialism began to transform institutions and landscapes. Among the fleeting beneficiaries were members of an intellectual elite, still able to rely on inherited wealth which enabled them to investigate at leisure the very processes of change which would soon render them obsolete as a social class. Among them was Thomas Pennant, who travelled widely in Britain and mainland Europe, establishing close contacts with leading contemporary natural scientists and philosophers, and publishing accounts of his tours.

1. Education, Life and Work

Pennant was born into four borderlands. Geographically, his home on his father's estate at Downing, Flintshire, North Wales, was situated in a transition zone between an apparently stable rural society and an increasingly dynamic urban milieu that was being forged out of incipient industrialization. Amidst tranquil farmsteads in the parishes of Whitford and Holywell, whose history he later chronicled, sprouted small-scale coal pits, leadmines and ironworks, serving not only immediate local demands for fuel and agricultural implements but also more distant markets via the canalized river Dee and its estuarine ports.

Socially, his family's ancestors had occupied the property, together with that of neighbouring Bychton, for several generations and were related by marriage to other members of the local gentry. As such, they were affluent, privileged and respected members of a regional, if not national, nexus whose status was rooted in ownership and exploitation of land. While still influential, customary relationships built on such foundations were beginning to be supplanted by alternative sources of wealth and employment, broadened personal horizons, and divergent power structures created by increasingly rapid economic change.

Culturally, for centuries Flintshire had been a veritable marchland, territory disputed between English and Welsh traditions, where medieval Anglo-Norman invasion from the adjacent palatine county of Cheshire had introduced different systems of land-holding, patterns of settlement, population distribution and language, and through which the incoming political and economic tide from the anglophone east had gradually eroded the indigenous Celtic substratum to the west. In a broader sense small localized communities were engulfed by that trend towards the creation of an enlarged world which enabled them to participate in transformations that were being experienced throughout western Europe and beyond.

Intellectually, too, the context of Pennant's life and work was one of ferment. As he entered manhood, and certainly by middle age, he was fully conscious of turbulent currents of thought in which critical scientific rationalism, arising from a combination of intensified natural curiosity and the material products of exploration and discovery, had begun to sweep away myth and superstition in favour, at least within Europe, of a more widely agreed view of the universe and its component physical elements. With it came a growing awareness of diverse human characteristics and relationships. During his later years radical political ideas, often engendered in Britain but seeding themselves more liberally overseas in France and America, once again disturbed the homogeneity of a dominant social group, opening up completely novel perspectives, the more exact outlines of which appeared during the nineteenth century.

In his own life Pennant personified the half-worlds which he inhabited. Having been nursed as an infant at the nearby farm cottage of Pentre and taught initially at home, he was formally schooled first at Wrexham under Rev. William Lewis and later in Fulham, London. Although precise details are not known and in his autobiographical writings he makes no claims to academic distinction, Pennant claimed to 'detest the unnatural education of schools', notwithstanding the fact that later both of his sons were to be sent away from home to such institutions. Whatever the content and quality of his own instruction, it is clear that he absorbed a thorough knowledge of Latin and French and acquired a profound appreciation of both English and Welsh history from extensive reading, notably of John Leland's *Itinerary in Wales*, William Camden's *Britannia*, Giraldus Cambrensis' *Descriptio Cambriae* and David Powel's *Historie of Cambria*.

By his own admission, it was to a gift of a book from his uncle, John Salusbury of Bachygraig, Tremeirchion, that he attributed his enduring fascination with natural history from the age of twelve. That volume was Francis Willughby's *Ornithology*, but over the next decade Pennant's interests broadened to embrace geology. At eighteen he matriculated at Queen's College, Oxford, but like many another young man of means and leisure did not take a degree. However, during one university vacation in 1746–7 he toured Cornwall and met the antiquarian-naturalist Dr William Borlase, an encounter that consolidated his predilection for studies of minerals and fossils. For more than twenty years, as Vicar of Ludgvan, Penzance, Borlase had been collecting material on natural and cultural phenomena throughout the county and in 1748 a meeting with Dr Charles Lyttelton, the newly appointed Dean of Exeter and future Bishop of Carlisle, led to the publication of some of his geological findings in the *Philosophical Transactions* and his election as a Fellow of the Royal Society in 1750. Pennant's subsequent career mirrors that of Borlase in several respects.

In the absence of a requirement to earn his own living, or as yet to become involved in the management of his father's estate and support a family, for the next decade Pennant was free to pursue his own interests and to travel. His first extended journey was undertaken in the summer of 1754 when he toured 'the hospitable kingdom of Ireland' on an itinerary that included Dublin, the Giant's Causeway, Coleraine, Londonderry, County Donegal, Strabane, Enniskillen, Galway, Limerick, Killarney, Kinsale, Cork, Cashel, Waterford and Kilkenny. However, such was the conviviality that his journal 'proved as *maigre* as the entertainment was *gras*', with the result that his observations were never published and readers who might have been interested had to await the accounts and insights of Arthur Young nearly a quarter of a century later.

Having married Elizabeth Falconer of Chester in 1759, Pennant settled to a busy domestic life at Downing, which he inherited on the death of his father in 1763, but his wife also died the following summer, leaving him to care for their two young children, Arabella and David. Despite these grievous losses, or possibly even in reaction to them, in February 1765 he set off from Dover on a continental tour that lasted exactly five months. Unlike other devotees of the 'Grand Tour' that was in vogue among more affluent members of contemporary English society, Pennant's purpose was not merely to wonder at the elegance of European cities or to reflect on foreign artistic achievements and quaint customs; least of all did he wish to assimilate the latest superficial fashion in dress or manners. On the contrary, the main objective of his excursion was to learn of recent empirical discoveries and philosophical ideas from eminent scholars, to examine museum collections and consult libraries, and to collect original scientific specimens or drawings of them. His route through France, the Swiss Confederation, the German states and the Low Countries was therefore directed towards Paris, Geneva, Berne, Zurich, Leiden and the Hague, in order to meet Comte de Buffon, Voltaire, Baron Albrecht Haller, Johann Gesner, L.T. Gronovius and P.S. Pallas (see *Geographers*, Volume 17), respectively. Subsequently, in October 1780 Pallas offered to promote French and German translations of Pennant's *History of Quadrupeds* and sounded him out as to whether he would like to become a member of the Russian Academy, but had to disappoint him the following year because the director's absence on an expedition meant that no public assembly was convened to elect foreign members.

Having begun *The British Zoology* four years earlier, on his return from the continent Pennant resumed that research and published the first folio volume on quadrupeds and birds, an achievement that led to his election to Fellowship of the Royal Society in 1767. That distinction brought him into even closer association with the foremost European scientists and in particular with the Society's future President and explorer of Newfoundland and Labrador, Joseph Banks, whom Pennant guided through the Vale of Clwyd during his tour of Wales later that year and whom he visited at his home, Revesley Abbey, Lincolnshire, the following May. With his reputation already established in learned circles, like Borlase before him, in 1771 he was awarded an honorary doctorate by the University of Oxford.

So committed was Pennant to the task of disseminating newly acquired visual information that from the spring of 1769 he employed in his household a young self-taught artist from Caernarvonshire, Moses Griffith, who became his inseparable companion on all his tours with the sole exception of that to Scotland later the same year. While Griffith worked virtually full-time for him over the next 30 years and produced more than 700 drawings for his books, further items were also commissioned from John Ingleby of Halkin and Paul Sandby.

The bookseller Benjamin White, whose more widely-known brother Gilbert addressed the letters comprising the first part of his *Natural History and Antiquities of Selborne* to Pennant, proposed a new two-volume octavo edition of *The British Zoology* and a revised four-volume edition was produced by 1770, illustrated with scores of engravings based on drawings by Griffith. Meanwhile, as was characteristic of his working practice, Pennant had been busy simultaneously completing his *Indian Zoology* and *History of Quadrupeds*. The former, begun in May 1769 in a desire to 'relieve my pen by the pleasure of the novelty and variety of the subjects' by 'forming a zoology of some distant country', was induced by another contact, namely that with J.G. Loten, who as Governor of Ceylon and the Dutch Indian Ocean islands had similarly employed artists to make drawings of exotic tropical animals and birds. These prints were transmitted in turn to Johann Reinhold Forster, who, with his son Georg, had accompanied Captain James Cook (this volume) on his second voyage, and who added an essay on the climate and soils of India before publishing the collection in Halle in 1781.

The *History of Quadrupeds* (1781) emerged from *Synopsis of Quadrupeds* (1771), in the preface to which Pennant maintained that originally it had been 'intended for private amusement, and as an Index, for the more ready turning to any particular animal in the voluminous list of quadrupeds by M. de Buffon', the title having been altered to accommodate vast additions. There followed *Arctic Zoology*, dedicated to the Royal Academies of Sciences of Stockholm, Uppsala, Lund and Trondheim, which had been initiated as a sketch of the zoology of North America, 'a work began long since when the British Empire was entire and possessed of the northern part of the New World'. Despite widespread co-operation received from his network of correspondents, Pennant evidently judged that the loss of the colonies as a consequence of the American War of Independence prejudiced any immediate prospect of attaining that objective, with the result that 'ages must pass before new colonization can push its westward progress to the space between the Atlantic and Pacific', and he accepted that a description of that territory must be left to the Americans themselves. As an alternative, in these volumes he confined his attention to the circumpolar regions and incorporated material on Europe and Asia, relying heavily on information provided by Banks, William Coxe, Forster, Pallas (from Georg Wilhelm Steller's collections and manuscripts), E.A.W. Zimmerman and other naturalists.

If some political upheavals in the eighteenth century curbed freedom of movement and access to information, elsewhere renewed peace and administrative stability opened up different travel possibilities. Following publication of *The British Zoology*, and perhaps aware of its shortcomings, in June 1771 Pennant resolved to rectify gaps in his knowledge by venturing 'on a journey to the remotest part of North Britain, a country almost as little known to its southern brethren as Kamschatka'. He sought to demonstrate that, after the turbulence of two Jacobite rebellions, Scotland could now be visited with safety and comparative ease, and to contribute to the reconciliation of the two peoples by presenting an objective account of the country's natural beauties, historic monuments, cultural traditions and recent economic developments. On this occasion he restricted his tour on horseback to the mainland, but returned with Revd John Lightfoot (who succeeded Daniel Solander as the dowager Duchess of Portland's librarian), the Gaelic-speaking Revd John Stuart and Moses Griffith in May 1772, when he was granted the freedom of Edinburgh and also sailed to the Hebrides. Publication of *A Tour in Scotland* was so well received, not least by the critic Dr Samuel Johnson, that the appearance of several editions in quick succession permitted inclusion of many new illustrations and the author implied, not without reason, that the book's popularity had caused the country to become '*inondée* with southern visitants'.

The years 1774–5 saw two journeys to the Isle of Man, the first in the company of his neighbour Paul Panton of Plas Gwyn and Revd Hugh Davies, author of *Welsh Botanology ... A Systematic Catalogue of the Native Plants of Anglesey* (London, 1813), who in 1790 became a Fellow of the Linnaean Society and helped Pennant with his *Indian Zoology*. Although a second visit by Davies identified 550 plants on the Isle of Man, no account of the tour was ever published, since Pennant's original journal and notes were lost.

On New Year's Eve 1787 Pennant lamented to a friend: 'I grow old and cannot support the plague of publication', and in his *Literary Life* (1793), when he had ceased to write for a public audience, he revealed that after his second marriage, to Anne the daughter of his neighbour Sir Thomas Mostyn in January 1777, he had lost his 'spirit of rambling'. Yet revised editions of earlier books and other 'home-travels' throughout southern Britain continued to appear. While they conformed to the literary device of conventional itineraries, strictly these were not accounts of single experiences; rather, they represented amalgams of several journeys made at various times. The only two to be published in Pennant's lifetime were *A Tour in Wales* (1778), to which was added in 1781 a second volume, originally titled *The Journey to Snowdon*, and *The Journey from Chester to London* in 1782. Having formerly been a frequent visitor to the capital for social and publishing reasons, Pennant had long grown disillusioned with the city's scandals, intrigues, squalor and violence, but after many years of interrupted effort in the spring of 1790 he issued *Of London*, revised in 1791 as *Some Account of London*, which became a standard guidebook.

In keeping with his increasing age and new-found domesticity, compounded by occasional attacks of asthma, influenza and gout, preparation of two final, but quite contrasting, books occupied Pennant's remaining time and energy. As a metaphor of his deepening introversion – even in December 1789 he confided 'I write so much and think so much that I lose the habit of talking' – in April 1795 he completed *The History of the Parishes of Whiteford and Holywell*, essentially a summary of the history and geography of his native district in north-east Wales, centred on chronicles of his own and the more extensive Mostyn estates.

Just as that volume went to press Pennant agreed terms with Benjamin White for the publication of the first instalment of *Outlines of the Globe*, without doubt his grandest concept and his most remarkable achievement. The precise origin of the scheme is not known but in a very real sense it represents the culmination of his life's activity. At the conclusion of the *Literary Life*, after noting the appearance of a second edition of his *Indian Zoology*, he observed: 'A few years ago I grew fond of imaginary tours and determined on one to climes more suited to my years, more genial, than that to the frozen north ... a voyage to India, formed exactly on the plan of the Introduction to the *Arctic Zoology*; which commences at such parts of the north as are accessible to mortals.' Although the whole compendium was conceived to comprise fourteen volumes (22 manuscript folio volumes), it appears that Pennant had little intention of publishing them and professed that it was only at the instigation of friends that he prepared for printing those sections dealing with the Indian subcontinent, the South-East Asian archipelago, and 'the miracle of governments', China, a task completed by his elder son in 1800. David Pennant also published three works that had been finished in manuscript before his father's death: *A Journey from London to the Isle of Wight* (1801), *A Tour from Downing to Alston-Moor* (1801) and *A Tour from Alston-Moor to Harrowgate, and Brimham Crags* (1804). The second and third date in reality from excursions made in 1773, and the first from May 1787 when David, who was soon to embark on his own third continental tour, accompanied his father.

A conservative political outlook, which might have been expected from inherited status and personal family linkages, was in Thomas Pennant's case mellowed by an openness of spirit and private generosity that may be attributed to the necessary breadth of his social acquaintances, to knowledge derived from direct experience, and to a scholarly curiosity into the nature of a wider world. Always aware of the fine divide between crippling misfortune and comfortable survival, his donations to less prosperous friends and institutions were frequent and unconditional. At his own expense he published Lightfoot's *Flora Scotica* (1778), Charles Cordiner's *Antiquities and Scenery of the North of Scotland* (1776), and had wished to do the same for George Low's survey of the Orkney and Shetland islands. Of equal concern in later years were practical issues that he believed arose from institutionalized injustices, such as the requirement for the poor to repair roads that mainly served the convenience of the rich, since mail coaches were exempt from tolls; burdensome taxation for the support of the militia; and restrictive legislation against local tradesmen. On these and other topics of principle he wrote pamphlets and petitions in which humane causes were argued with measured reason.

He was never enthusiastic in the cause of democracy, and the course and outcome of the war in America and the barbarous excesses of the French Revolution intensified his fear of potential civil unrest in Britain fostered by poverty, political corruption and radical ideologies. In life, as in his scholarly work, ambivalence dogged his thought. In July 1788, for example, he bemoaned the plight of the cotton trade and its impact on manufacturers in nearby Holywell; only six weeks later he confronted the horror of the underlying trade in human beings as he 'clambered over the slave ships' during a 'jaunt to Liverpool', yet the close connection between the two did not appear to be realized.

Having received numerous honours from learned societies in Britain, Europe and America, Pennant remained anxious to be of service to his local community. Though he professed to abhor committees and associations, he fulfilled his duties as a magistrate conscientiously, promoted road improvements, imported grain for the poor in times of shortage, and initiated charitable subscriptions in aid of bereaved families of friends, servants and tenants. When Britain seemed to be threatened by invasion from France he patriotically revived bounties payable to seamen recruited to defend the state.

Disenchantment with the conduct of public affairs at national level, compounded by increasing physical infirmity, led him to withdraw from London society during the last ten years of his life. Saddened by the loss in quick succession of his sister and several old friends, and disapproving of the marriage of his elder daughter Arabella, he and his wife were devastated by the death of their younger daughter Sarah at the age of fourteen. Thomas Pennant himself died at Downing on 16 December 1798.

2. Scientific Ideas and Geographical Thought

Introducing the 1766 edition of *The British Zoology*, Pennant wrote: 'The one great Object of this History is to Promote the Glory of the Almighty by demonstrating his Wisdom in the Works of the Creation; the other, to relieve the Indigent, the Orphan, the Deserted of our own Country', the former being phrased in almost identical terms to the title of Ray's book, the latter being a reference to the assignment of any profit from the volume to the Welsh Charity School in Clerkenwell, London. The preface to the fourth edition (1776) broadened this view, asserting that 'the purpose to exalt our veneration to the Almighty is the principal end of this sublime science' which should be used to the benefits of human society. Regarding zoology as 'the noblest part of natural history', he proceeded to elucidate the links between food plants and dietetics as necessary to medicine and the eradication of disease, extending the reasoning to relate the parallel contribution of natural products to human comfort through clothing, the manufacture of which in turn provided employment and underpinned a country's domestic economy. His readers were exhorted to pursue a knowledge of natural history in the open air, that is by field observation, thereby promoting 'health of body and cheerful contentment of mind'.

From Pennant's writings it is evident that his scientific approach to natural phenomena was conditioned by those of three great scholars – John Ray, Carl Linnaeus and Georges-Louis Leclerc (Comte de Buffon). Of these his deepest admiration was reserved for Ray, whose friendship and collaboration with Willughby a century earlier set the mould for Pennant's own work. Accepting that 'the common fate of antiquated systems' is to be consigned to oblivion, his devotion to Ray's *Synopsis* as the basis for his own *History of Quadrupeds* was qualified by a belief that it was constrained by its period, above all by the limited extent of commerce and attendant sources of information in that bygone era. Even in his own day, Pennant admitted, ignorance of many links in the chain of beings remained so great that perfection could not be expected. As a step towards that goal, allying himself more with Buffon than Linnaeus, he adopted a stance to which few would now demur, namely that classifications should be multivariate: 'We ought ... to drop all thoughts of forming a system of quadrupeds from the character of a single part; but if we take combined characters, of parts, manners, and food we bid much fairer for producing an intelligible system, which ought to be the sum of our aim'. His *Arctic Zoology, British Zoology* and *History of Quadrupeds* were all framed by a system based on five Classes (quadrupeds, birds, reptiles, fish and insects), each composed of Divisions or Orders, Genera and Species. Ignoring anatomical details, each genus was named, with synonyms in several languages, described, and its uses and natural history set out. In the introduction to *Arctic Zoology*, a 'knowledge of the geography, climate and soil and general view of the production of the countries' embraced by the survey assumed no more significant role than that of a necessary preface, but the text demonstrates a much deeper concern for the totality of these integral relationships, not least spatial ones. Special acknowledgement was given to Zimmerman's *Specimen Zoologiae Geographicae Quadrupedum* (Leiden, 1777) which featured a map 'with animals plotted in its proper climate in a manner perfectly new and instructive'. Of vital importance to Pennant was consideration of the causes of regional differences in the density and diversity of fauna, citing the contrasting extremes of Spitzbergen and Senegal. Moreover, a crucial amendment to the second edition in 1792 was the inclusion of two maps by William Palmer on which various Siberian peoples and North American Indian tribes were also located.

In mid-February 1755 he began a correspondence with Linnaeus which was to continue for seventeen years and involved an exchange of ideas, publications and specimens. So impressed was Linnaeus that in 1757 he was instrumental in having Pennant elected to membership of the Royal Society of Uppsala. Two explicit motives had lain behind the writing of *The British Zoology*, a reluctance 'to let illustrious foreigners monopolise the field' in the European revival in the study of natural history, and a desire to present another concise summary of phenomena in one country on the model of Linnaeus' *Fauna Suecica*, which had been published twenty years earlier in 1746. Meanwhile, the latter's seminal *Systema Naturae* had appeared, offering a conceptual underpinning for future work which rivalled that of Ray. Temperamentally disinclined or academically ill-equipped to compete with Linnaeus' sustained logic and specialist command of detail, Pennant contented himself with marginal points of disagreement, while praising the monumental grandeur of the scheme.

The highlight of Pennant's continental tour was almost certainly the hospitality he received in Paris from the Comte de Buffon, who provided introductions to other collectors of natural curiosities and to illustrators, and indeed gave him many specimens and prints. Buffon's stylish and all-embracing *Histoire naturelle* (1749) had cast doubt on the literal truth of a Biblical flood, had sought to argue for longer gradual processes of physical change instead of catastrophism, criticized Linnaean classification as artificial and mechanical, and was itself pilloried for appearing to contradict theological orthodoxies on the history of the earth. Pennant's admiration went beyond the linguistic elegance of Buffon's works and evidently the two warmly appreciated each other's studies, though occasionally Pennant quibbled over acknowledgements which he felt were his due.

The name of one other great naturalist recurs in Pennant's contributions to scientific thought. Having met him in 1765, Pennant had suggested to P.S. Pallas that the latter should write a history of quadrupeds on a modified system of Ray's classification and received his outline in January 1766. Unfortunately, the project lapsed when Pallas returned to Berlin and accepted an invitation to participate in the Russian Academy of Sciences' expedition to Siberia (1768–74). Surviving correspondence dating from 1766 to October 1781 confirms regular contact between Downing and St Petersburg, including substantial exchanges of mineral ores, fossils, plant seeds, specimens of birds, recent maps, books, together with corrections, clarifications and advice on publication matters, and observations on contemporary events and social life in Russia. Their collaboration also exemplifies how Pennant's views of others were modified by shifts in continental politics. In *History of Quadrupeds* (1781) he had applauded Catherine the Great's 'munificent' patronage of Pallas, and in acknowledging his debt to him in the preface to *Arctic Zoology* (1788) he praised her as the 'illustrious empress of Russia'. By April 1791 she was branded 'that dreadful woman of the north' for slaughtering 30,000 'gallant Turks', while in January 1793, during intensified reaction against the terror in revolutionary France, he still expressed distaste for her as Britain's new ally, only conceding that 'to deal with monsters what can you select more proper than a monster'.

When the nineteenth-century British prime minister Benjamin Disraeli was asked, somewhat provocatively, for his views on the continued existence of an established church in England, he is reported to have replied that at least it had the merit of placing one gentleman in every parish. The role of the Anglican, or in Scotland the Presbyterian, clergy, or indeed of members of the medical profession, in amateur research in natural science has yet to be fully documented, but the substantial contributions of many were recognized in the letters and published works of Thomas Pennant. The extent to which he shared their theology or *Weltanschauung* may be debated. Certainly he declared himself to be 'a sincere well-wisher to the pure form of the church of England' and expressed disdain for radical nonconformist Methodism, traits of which he detected in his artist Moses Griffith in 1793. It would, however, be simplistic to attribute to Pennant's alert mind an uncritical acceptance of established religion or to regard his biological views as mere parallels to the rigid hierarchy in the prevailing ecclesiastical system.

On one issue, the age of the earth and the duration of life on its surface, his knowledge of fossils did not seem to undermine his belief in the Biblical version of the Creation, nor acceptance of James Ussher's calculations of geochronology on the basis of genealogies in the Book of Genesis. For instance, in *Outlines of the Globe* he recounted Du Halde's legendary tale of lands in China reclaimed from the sea following the Noah flood. Noting that the Emperor Yau reputedly ruled from 2237 BC, Pennant merely commented that 'the period in which he was born could not therefore have been long after the universal deluge', an event dated by Ussher to 2349 BC. A similar adherence to catastrophism is found in *Arctic Zoology*: in beginning his narrative at the Straits of Dover he affirmed that 'no certain cause can be given for the mighty convulsion which tore us from the continent'. Leaving no room to doubt that the corresponding chalk strata in England and France were formerly united, as an alternative to seeing a 'once peninsulated Britain' rent aside by earthquake, however, he did raise the possibility that the isthmus may have been 'worn through by continual clashing of waters'. It may be no coincidence that Pennant's first paper in the *Philosophical Transactions* was an account of the earthquake that occurred near Downing in 1750. *A Tour in Wales* hints at vulcanism and tectonic movements in shaping Snowdonia's mountains, where discovery of shells in the top-most rocks suggested to him that Glyder Fach was 'a sort of wreck of nature, formed and flung up by some mighty internal convulsion'. Regrettably, the fundamental views of his almost exact contemporary, the Scottish geologist James Hutton, may not have been available to him, since the first paper on the *Theory of the Earth* was not published in Edinburgh until 1788 and the complete book not until 1795.

Concessions to a long evolutionary course became more evident when Pennant was discussing human and animal characteristics and behaviour. The whole question of cultural change was treated with greater awareness of gradual development which clearly derives from his studies of written history and of archaeological remains. Endogenous forces were recognized as being capable of achieving progress, but the primary thrust of *Outlines of the Globe* was that population migration and cultural transmission were more potent, a view entirely in harmony with the implicit philosophy of the colonial era. In an intriguing foretaste of Darwin, the role of islands as stepping-stones in the diffusion of people, plants and animals had already been presented in *Arctic Zoology*, but the notion was given more vivid and varied exemplification in the context of interaction between Chinese, Japanese and European peoples in South-East Asia. Dispersion of anthropological types, languages, religious beliefs, laws, rituals and customs, and food crops (mulberry, grape, lemon, sugar cane, tea, coffee, rhubarb), all find illustration of principles that were unfortunately never explicitly expounded. Underlying a fascinating range of contacts, not least the deliberate experimental transfer of plants by returning explorers, there remained a residual unquestioned belief in a single aboriginal hearth of civilization that stemmed from Judaeo-Christian legend.

In *Arctic Zoology* he accepted that different manners, customs and physical appearance identified in remote contemporary peoples derived from cultures shared by our common ancestors, thus permitting comparative studies of artefacts found in contrasting parts of the world. At the same time, distinctions made between groups of biological creatures were not invariably free from value judgements. Dissenting from Linnaeus and regarding primates as the 'chiefs of creation', in his own words 'vanity will not suffer me to rank mankind with apes, monkeys, mancaucos and bats'. Similarly, he refused to countenance a Linnaean Order in which 'the most intelligent – the elephant – is made to associate with the most discordant and stupid of the creation, with sloths, ant-eaters, and armadillos, or with manatees and walruses, inhabitants of another element'. The criticism thus became confused and unscientific by invoking diverse and unrelated criteria, namely supposed intelligence and habitat. An appendix to the *Literary Life* (1793) devoted to the inhabitants of Patagonia – coincidentally, an area settled by Welsh colonists in the 1860s – dwelt at length on anthropological and cultural differences between groups of pre-Hispanic peoples, differentiating between them qualitatively on scales of physical stature, miscegenation, economic life, technology, customs, and spiritual beliefs. Although the account was rendered with all gravity, at times the essay almost exuded the allegorical flavour of Jonathan Swift's *Gulliver's Travels*, a reference to which did in fact occur in *Outlines of the Globe* when Pennant was discussing Japan.

Again, in a parallel sphere, while postulating possible variation in animals as a result of adaptation to climatic differences, Pennant uncritically followed Montesquieu in tracing perceived human psychological traits to environmental causes. At a very basic level he considered the climate of the 'Torrid Zone' to be directly responsible for enervation of body and mind, when 'every idea is too often lost in irresistible indolence', while attributing to the sun's heat the ferocity of its wild animals, human barbarism, and even the 'wonderful extravagance in Indian mythology'. In terms redolent of European superiority, he dismissed the religion of the Japanese as idolatry and deplored as obscene sculptured images of some Hindu deities. By contrast, at home in Britain local superstition which associated prehistoric burial sites with 'druidical' rituals was recorded virtually as historical fact.

Analysed in detail, his later writings do reveal a comprehensive, if imprecisely articulated, and even sensitive appreciation of interacting processes, an acute sense of the significance of distribution or relative location, and the value of interdisciplinary syntheses that cut across narrow artificial specialisms. Their ultimate weakness was that none of these views found coherent systematic expression in print.

Five primary and secondary sources may be distinguished in Pennant's methods of acquiring empirical information about the world around him. First, he spared no effort in scouring previously published research. In sections of *Outlines of the Globe* which dealt with India, China and South-East Asia he demonstrated a familiarity with the writings of, *inter alios*, Al-Edrisi, Herodotus, Lucan, Pliny, Pompey and Strabo, and assimilated the discoveries of William of Ruysbruk, Marco Polo (see *Geographers*, Volume 15), and Samuel Purchas, not to mention the fantasies of Sir John Mandeville, together with more realistic contemporary travel accounts by Bernier, Coxe, Dampier, Forster, Kaempfer, Le Poivre, William Roxburgh, Patrick Russell and Thunberg.

Second, while on his tours, acute field observations recorded in a journal represented a significant fund from which material was later selected for publication. Occasionally, immediate gathering of facts or anecdotes remained incomplete and gaps were left in preliminary notes, for example for subsequent verification of place names. However, as an implied accusation of haste and superficiality, Horace Walpole's deprecating comment that 'he picks up his knowledge as he rides' was only partially valid: indeed, given the pace of travel on horseback in pre-industrial Europe, it may even be considered entirely appropriate. While he sometimes shows undue deference to aristocratic owners of prominent houses on his route, his artless ease in company enabled him to strike up a rapport with humble folk, from whom he teased out details of their daily routine and beliefs.

Third, both before and after specific journeys Pennant went to extraordinary lengths to obtain additional data which might be of value or interest. Although his first Scottish tour seems to have been embarked upon with little preparation, the second was foreshadowed by an open letter to ministers of religion seeking responses to a score of queries concerning 'objects most worthy the observation of a traveller', with an injunction to them to collect items indicated. This attempt to collate standardized evidence on a limited list of topics, in addition to details of events, personalities or features of purely local importance, recalls the procedure of parochial enquiries adopted by Pennant's fellow countryman, the naturalist, antiquary and Keeper of the Ashmolean Museum at Oxford, Edward Lhuyd. Lhuyd, whose manuscripts Pennant had used in compiling material for his *Tour in Wales*, had himself travelled through parts of Ireland and Scotland, and had also been a close colleague of Ray. Pennant's web of correspondents, meticulously constructed over many years, stretched from Russia to North America and from Scandinavia to Gibraltar, and all served him admirably, as he did them.

Fourth, this practice was reinforced by the use of local guides. Just as he acted in that capacity for those who visited him in Wales, when in unfamiliar surroundings he took advantage of the expertise of the Gaelic scholar John Stuart during the tour of the Hebrides, and of similar skills from two other clergymen, the Anglesey botanist Hugh Davies and the Flintshire antiquarian John Lloyd, who corrected his interpretations of Welsh terms and place names and gave freely of their encyclopaedic knowledge of regional history and popular customs. Such intermediaries narrowed the potential gulf between the observer and the observed, thereby reducing the scope for misunderstanding or error. Nevertheless, a paradox that he might have accepted reluctantly was that, especially in Wales and Scotland, he often sought information from that very social group that, through education and anglicization, had lost any empathy it may have once shared with the majority of the common people.

Finally, among the most distinctive features of Pennant's published works is the inclusion of numerous fine, often hand-coloured, engravings by Peter Mazell of drawings, prints, coats of arms and maps. For example, the *Outlines* volumes referred to above incorporated prints of flora and fauna culled from John Latham, architectural and landscape views by Thomas Daniell and William Hodges, maps by Aaron Arrowsmith, Alexander Dalrymple and James Rennell (this series, Volume 1), all serving the two-fold purpose of adding attractive illustration to the text and providing supporting visual evidence of phenomena that would have been totally unfamiliar to incredulous non-specialist readers. For that reason alone he deserves to be remembered as an outstanding collector of antiquities and curiosities as well as a bibliophile.

In this regard the sequential processes of European imperial expansion were particularly fruitful: British military campaigns in India emphasized encounters not only with indigenous peoples, but also with Portuguese and Dutch rivals for control of the Indian Ocean; strategic expeditions inland from coastal bridgeheads resulted in improved knowledge of strange creatures, cultures and environments; foundation of colonies and 'manufactories' tapped resources for trans-oceanic trade, linking the commonplace with the exotic in the daily lives of both communities. Yet even at the end of the century it is noticeable how ports remained dominant in the pattern of towns plotted on the most recent maps of India. As Pennant attempted to chronicle events impartially for a partisan British public divided by the controversy over India's governor Warren Hastings, how often he must have relived his experiences when, towards the close of his continental tour in 1765, in the Hague and Amsterdam he had come upon stores of oriental treasures carried half-way round the globe by Dutch merchants.

3. Influence and Spread of Ideas

The craft nature of printing technology in the second half of the eighteenth century meant that mass production of books, journals and newspapers, and the more extended literacy that it fostered, still lay several generations in the future. Nevertheless, publications such as those by Thomas Pennant, whether representing specialist science or directed towards a broader readership, became relatively widely known through the intellectual networks created by personal contacts in a fashionable social elite or through organized gatherings of those sharing more serious interests in new discoveries or ideas.

Pennant's contributions to taxonomy may be considered very modest when set alongside the cathedral-like edifices of Ray, Linnaeus and Buffon. He did not construct an original alternative but merely added adjustments to theirs. However, in doing so he took the process of classification beyond the first stage of formal description by using extra criteria that may be called functional, or even utilitarian from an anthropocentric point of view, i.e. those indicating the value of plants and animals in the service of mankind. As yet, neither spatio-ecological linkages nor evolutionary mechanisms were fully understood and such parameters had to await the next generation.

As Pennant perceived in the preface to *The British Zoology* (1776), 'the generality of mankind rest contented with ignorance of their native soil, while a passion for novelty attracts them to a superficial examination of the wonders of Mexico or Japan'. He regarded it as his duty to deepen that transient public interest and shallow acquaintance with distant regions, while simultaneously stimulating a popular appetite for accounts of places equally fascinating but less remote. It was this diversion into accounts of travels within the British Isles that ensured his long-term reputation and established a model that was soon taken up by many other writers. Even the characterization of Pennant as 'the father of Cambrian tourists' minimizes his importance, for not only did he help to rediscover regions other than Wales, but he also pioneered a new genre in literature in which personal experience during the course of purposeful travel was blended with encouragement to probe further, to extend or step aside from the itineraries of his own *Tours*. That in itself inevitably resulted in promoting awareness of heritage and documenting newly observed natural phenomena, and on the eve of the world's most explosive transport revolution its impact cannot be underestimated.

The particular significance of the tours of the continent, Wales, Scotland and northern England lies, first, in their description of pre-industrial economies and often pre-literate societies at home and abroad. Diversity in rural scenes always caught his attention, from the shepherds of Artois to the goatherds of Snowdon. On the river Seine his boat met rafts of timber sent from Burgundy to fuel Paris, and in the beech forests around Buffon's home at Montbard he saw clog-makers, coopers and charcoal-burners at work. Transhumance and the traditional making of butter and cheese was just as visible on the summer grazings of upland Wales and the Grampian Mountains as in the *Sennerei* of the Swiss Alps. Pastoral views of Chantilly and Auxois contrasted sharply with the 'black and dreary waste' of Delamere Forest, the 'vast nakedness' of the treeless Yorkshire Wolds, the barren flatness of the Lincolnshire Fens. To shelter from the rain in a Perthshire fisherman's cottage or to view an August dawn rise like a theatre curtain over Anglesey proved just as intriguing as an inspection of Haarlem's flower gardens or vineyards along the Rhine.

Second, the *Tours* depict transformations then taking place. Some removed from the landscape the last vestiges of archaic ways of life: others created the material foundations of modern communities that were to endure for at least another century. Well in advance of the surveys of Arthur Young and the Board of Agriculture in Britain, Pennant reported agrarian improvements, enclosure of arable land, reclamation of bogs for experimental cultivation, schemes for planting trees on bare hills. He saw how internal migration relocated ability and energy when poverty and expectation drove peasants from Savoy to harvest work in Burgundy or more permanently to Paris, and Scots Highlanders to England and America from their windowless, broom-thatched cottages which he deplored as 'the disgrace of North Britain, as its lakes and rivers are its glory'. Alongside tales of folklore, such as the superstitious use of mountain ash and honeysuckle to protect cattle from witches in Banffshire, he noted that inoculation against smallpox was already being practised further north in Caithness, and entertained the hope that health cures of which he partook at spas and resorts like Bath, Brighton and Buxton would ultimately become accessible to all those in need. Industrialization and urbanization, represented by references to street lighting, paving and cleansing in Paris and London, and employment in porcelain factories at Vincennes and Sèvres, Gobelins tapestry works, silk mills in Macclesfield and Lyons, were replacing domestic crafts and modifying squalid rural lifestyles, at least for a minority. Colonialism and global commerce as foundations for European manufacturing had begun to flourish, while new internal resources were being tapped: around Glasgow, 'the best built of any modern second-rate city I ever saw', breweries and collieries supplied Ireland and America, and raw materials for tobacco, sugar and linen industries were imported, while Paisley was set to rival Manchester in production of linen and cotton goods. Further east in central Scotland Carron ironworks, founded on a greenfield site only a dozen years earlier, had become the greatest in Europe. Coastal shipping off Northumbria was diversifying from fishing to transporting fuel, now able to penetrate to inland towns via newly constructed canals and improved river navigation. In the English Lake District, relatively unfrequented before Wordsworth's birth, peat would soon be superseded on domestic hearths by coal from Lancashire. Pennant's continental tours took him from dirty narrow streets in medieval walled towns whose gates still closed at dusk to a view of a forest of ships' masts in Amsterdam's bustling port, symbol of the new open economic age and destination of immense wealth brought from as far afield as Greenland whaling grounds and East Indies plantations.

Everywhere from Snowdonia and the Scottish Highlands to Dauphiné, the Jura and Pays de Vaud, Pennant depicted the mountain backdrop in the same timorous vocabulary beloved of the Romanticists – 'frightful mountains', 'horrid torrents', 'deep narrow chasms between vast cliffs', 'the chilling and horrible prospect' of Swiss glaciers, or the 'awefully magnificent' pass above Birnam Wood on the river Tay. Although in some ways presaging the travels of landscape poets and essayists such as Coleridge, Shelley, Southey, and Wordsworth, Pennant's accounts were more solidly grounded in objective reality, appreciative of the picturesque but never blinkered against what they might decry as man-made and ugly. Within his own limitations he attempted to provide a rounded portrayal of places he visited. Arguably, his background in natural science enabled him to see the complete geography, and he may be considered as a forerunner both of regional 'tourists' such as Arthur Aikin, William Bingley, George Borrow, William Gilpin, B.H. Malkin, Thomas Roscoe and Richard Warner, and also of guidebook publishers like Karl Baedeker or Samuel Lewis with his *Topographical Dictionary*. Across the Atlantic, Benjamin Smith Barton, the distinguished professor of botany and medicine in Philadelphia, developed interaction between those two disciplines, as had been suggested in *The British Zoology*, and paid his old acquaintance the compliment of naming his son Thomas Pennant Barton.

When Thomas Pennant's writings are taken together, their range, literally from the parochial to the global, testifies not only to the scope of his curiosity but also to the breadth of his intellect, since all were pursued with the same vigour in the search for refined information or insight. Even more exceptional was the extent of his vision, that so much could be drawn within the compass of one person's knowledge that it was possible for him to present an eighteenth-century picture of the whole world.

Acknowledgements

The portrait of Thomas Pennant by Thomas Gainsborough (1776) is reproduced by permission of the National Library of Wales. I am also indebted to Alan J. Giddings of the Manuscripts Section at the National Maritime Museum, Greenwich, for details of Pennant's papers held there.

Bibliography and Sources

1. ARCHIVAL SOURCES

The main collection of manuscript letters and journals relating to Thomas Pennant, together with a complete collection of his published works, is held at the National Library of Wales, Aberystwyth, while the 22 bound manuscript volumes of *Outlines of the Globe* are preserved at the National Maritime Museum, Greenwich. Downing estate papers may be found at the Clwyd (Flintshire) Record Office, Hawarden; other correspondence exists at the National Library of Scotland (George Paton letters), and the Warwickshire County Record Office (Yorke-Pennant letters).

Watercolours, drawings and portraits by Moses Griffith are held by the National Library of Wales, the National Museum of Wales, Cardiff, and the Clwyd Record Office.

2. REFERENCES ON THOMAS PENNANT

Short summaries of Pennant's life and work appear in the *Dictionary of National Biography*, Vol. 15, 765–8, and *Nouvelle Biographie Générale* Vols. 39–40, Paris, 1863, 530–2, the latter based on Georges Cuvier's entry in *Biographie Universelle*, vol. 33. A biographical essay is also included in J.H. Parry, *The Cambrian Plutarch*, Simpkin & Marshall, London, 1824.

Bevan-Evans, M., 'Thomas Pennant and Downing', *Flintshire Historical Society Publications*, Vol. 14 (1953–4), 72–9.

Davies, Ellis, 'Thomas Pennant', *Journal of the Historical Society of the Church in Wales*, Vol. 2 (1950), 87–96.

de Beer, G.R. (ed.), *Thomas Pennant Esq.: Tour on the Continent 1765*, The Ray Society, London, 1948.

Evans, Ronald Paul, 'Thomas Pennant's writings on North Wales', unpublished MA thesis, University of Wales, 1985.

Evans, R.P., 'Thomas Pennant (1726–1798): "The Father of Cambrian tourists"', *Welsh History Review*, Vol. 13, No. 4 (1986–7), 395–417.

Moore, Donald, *Moses Griffith 1747–1819: Artist and Illustrator in the Service of Thomas Pennant*, Welsh Arts Council, Cardiff, 1979.

Powell, L.F., 'The tours of Thomas Pennant', *The Library: Transactions of the Bibliographical Society*, Vol. 19, No. 2 (1938), 131–54.

Price, C.J.L., 'Thomas Pennant', *Welsh Anvil*, Vol. 8 (1958), 59–73.

Rees, Eiluned and Walters, G., 'Pennant and the "pirates"', *National Library of Wales Journal*, Vol. 15, No. 4 (1967–8), 423–36.

Rees, Eiluned and Walters, G., 'The library of Thomas Pennant', *The Library: Transactions of the Bibliographical Society*, Vol. 25, No. 2 (1970), 136–49.

Urness, Carol, *A Naturalist in Russia: Letters from Peter Simon Pallas to Thomas Pennant*, University of Minnesota Press, Minneapolis, 1967.

Walters, Gwynfryn, 'The tourist and guide book literature of Wales 1770–1870', 2 vols, unpublished MSc thesis, University of Wales, 1966.

3. SELECTED BIBLIOGRAPHY OF WORKS BY THOMAS PENNANT

1750–5	'Of Some Fungitae and Other Curious Coralloid Fossil Bodies', *Philosophical Transactions of the Royal Society*, Vol. 10, 513–16.
1763–9	'On the Different Species of the Birds Called Pinguins', *Philosophical Transactions of the Royal Society*, Vol. 12, 91–100.
1766	*The British Zoology*, Vol. 1, J. & J. Marsh, London.
1768–70	*The British Zoology*, 4 vols revised, Benjamin White, London.
1769	*Indian Zoology*, Henry Hughes, London.
1770–6	'Account of Two New Tortoises', *Philosophical Transactions of the Royal Society*, Vol. 13, 266–73.
1771	*Synopsis of Quadrupeds*, John Monk, Chester.
1771	*A Tour in Scotland*, John Monk, Chester.
1773	*Genera of Birds*, Balfour & Smellie, Edinburgh.
1774	*A Tour in Scotland and Voyage to the Hebrides, 1772*, John Monk, Chester and London. (Second edition, Benjamin White, London, 1776.)
1774	*A Tour in Wales*, Henry Hughes, London. (Also two-volume edition, Benjamin White, London, 1784; a three-volume edition, *Tours in Wales*, was edited by Sir John Rhys, and also translated into Welsh as *Teithiau yn Nghymru*, Caernarfon, 1883.)
1781–5	'An Account of the Turkey', *Philosophical Transactions of the Royal Society*, Vol. 15, 67–82.
1781	*History of Quadrupeds*, 2 vols, Benjamin White, London. (Enlarged from *Synopsis*.)
1781	*The Journey to Snowdon*, Henry Hughes, London.
1782	*The Journey from Chester to London*, Benjamin White, London. (Pirated edition Dublin, 1783).

1784–7	*Arctic Zoology*, 2 vols, Robert Faulder, London. (A German translation was published in Leipzig in 1787 and a French one in Paris in 1789.)
1790	*Of London*, Robert Faulder, London. (Second edition, *Some Account of London*, 1791. A German translation appeared in Nuremberg in the same year.)
1793	*The Literary Life of the Late Thomas Pennant Esq., By Himself*, B. & J. White/Robert Faulder, London. (A version of this essay appeared in *Anthologia Hibernica*, August 1793, 83–93 and September 1793, 176–86.)
1796	*The History of the Parishes of Whiteford and Holywell*, B. & J. White, London.
1798–1800	*Outlines of the Globe*, 4 vols, Henry Hughes, London.
1801	*A Journey from London to the Isle of Wight*, 2 vols, Edward Harding, London.
1801	*A Tour from Downing to Alston-Moor*, Edward Harding, London.
1804	*A Tour from Alston-Moor to Harrowgate and Brimham Crags*, John Scott, London.

Chronology

1726	Born 14 June at Downing, Whitford, Wales
1744	Entered Queen's College, Oxford
1746–7	Visited Cornwall; met Dr William Borlase
1752	Climbed Snowdon
1754	Toured Ireland. Elected Fellow of the Society of Antiquaries
1755	Began correspondence with Carl Linnaeus
1757	Elected Member of the Royal Society of Uppsala
1759	Married Elizabeth Falconer (died June 1764)
1761	High Sheriff of Flintshire
1763	Succeeded to the Downing estate
1765	Toured western Europe; met Buffon, Voltaire and Pallas
1767	Elected Fellow of the Royal Society
1769	Toured Scotland
1771	Awarded LLD by the University of Oxford
1772	Second tour of Scotland and the Hebrides
1774	Visited the Isle of Man
1777	Married Anne Mostyn
1781	Elected Member of the Society of Antiquaries, Edinburgh

1784   Elected to the American Philosophical Society and to the Royal Academy of Sciences, Stockholm

1787   Toured south coast of England from Dover to Land's End

1798   Died at Downing, 16 December

Colin Thomas is Reader in Geography at the University of Ulster, Coleraine, Northern Ireland.

Charles-Eugène Perron

1837–1909

Peter Jud

Photo by G. Dajoz. Courtesy of Museum d'Histoire Naturelle, Geneva, Switzerland

There was a small coterie of those who were profoundly influenced by the great French geographer Elisée Reclus (1830–1905; this series, Volume 3). Indeed, it could be said that Charles Perron was one of those who were enabled to achieve something in science or in philosophy primarily through their acquaintance with this important writer and thinker at a particular critical point in their lives. Charles Perron was the cartographer of the *Nouvelle Géographie Universelle* (*N.G.U.*), the main work of Reclus. It was only through his contribution to the *Nouvelle Géographie Universelle* that Perron, a comparatively unimportant artist, became such a well-known cartographic illustrator, thus laying the foundations for his independent cartographic work in the years that followed.

1. Education, Life and Work

Charles-Eugène Perron was a descendant of a family that migrated to Switzerland from Savoy in the eighteenth century. He was born on 6 December 1837 in Petit-Saconnex, a suburb of Geneva, the son of a painter on enamel who at a later stage in his life became the director of a hospital. Little is known about Charles' youth, except that he obviously wanted to become an artist, as he studied at the Art High School in Geneva, where he was taught by Barthélemy Menn.

At the age of about twenty, Charles left Geneva to spend five years in Russia. He described his activities in that time as consisting of 'making portraits on enamel'. During this time he probably obtained a good insight into the political and social conditions of Tsarist Russia: he may also have been in touch with some of the young adherents of nihilism.

About the years that followed, Perron provides this information: in 1862 he, together with his father Georges, published an album of coloured plates with the title *Armée suisse, types militaires, dessinés par Ch. Perron*, which was reasonably successful, and was followed by a few other individual plates. Then he moved to La

Chaux-de-Fonds (in the Swiss Jura), where he did industrial enamel painting. Back in Geneva he opened a studio, which continued to operate until 1870; he then spent two winters in Menton (France) involved in photographic work.

After his return from Russia he began to be associated with Geneva socialists. He became a busy member of the Geneva section of the Association Internationale des Travailleurs. In the second half of the 1860s, he was a close associate of the revolutionary and anarchist Mikhail Bakunin (1814–1876), who lived first on the upper shore of Lake Geneva, then in Geneva itself. As editor of a number of different anarchist periodicals (for example, *L'Egalité, Le Travailleur*), Perron was actively involved in spreading the socialist message. It was probably through his work as an editor that he met Elisée Reclus in Paris in December 1869, when Reclus promised to contribute to *L'Egalité*.

From the 1870s, Perron withdrew from the Internationale movement, although he did not renounce his convictions. He became sceptical of the political and philosophical radicalism of Reclus and Bakunin; perhaps he was even afraid of it, and he came to believe that he was wiser, more realistic and more practical than some of his comrades. Max Nettlau, the historian of the Anarchist movement, passes on the opinion of the Russian Peter Kropotkin (1842–1921; see *Geographers*, Volume 7), who described Perron as one of those 'who criticize, criticize and criticize again without doing anything'. His relationship with Reclus was affected by the sharp contrast in the characters of the two friends. Between vivacious Reclus and the slow, somewhat stubborn Perron a really deep relationship never developed, even after long years of collaboration.

Their professional collaboration began in 1874, when Reclus moved from Tessin (in the southern part of Switzerland) to Lake Geneva. Reclus had been working on the *N.G.U.* since his arrival in Switzerland in 1872, and although Perron had probably never drawn a map before, he entrusted to him the cartographic illustrations for the work. (The *N.G.U.* was published in nineteen volumes between 1876 and 1894. Every volume was published at the end of a year, and is dated the following year.) All the maps in the first volume, and a large majority of those in the second, were drawn by the cartographers of the publishing house Hachette & Cie, Paris. This arrangement did not suit Reclus, as it meant posting maps back and forth during correction, and the situation was improved when Reclus was able to supervise Perron's work close at hand. From volume 2 onwards a growing proportion of the maps were made by Perron. From Volume 6 onwards all the single-colour maps, and from Volume 10 onwards all the multi-coloured maps, are signed by Perron. As he grew into the task, so his professional expertise improved. The total production is quite remarkable: up to 1893 Perron completed 2700 single-coloured maps in the text, plus about 60 larger, multi-coloured maps of varying size. Perron's contribution to the *N.G.U.* is remarkable both for the number of the maps, and for their uniformity, instructive nature and usefulness. Reclus directed the work firmly: he decided on the selection of maps to be included, the regions represented, etc. His text was lively and instructive, yet thoroughly scientific, and the maps had to be subordinate to it. Perron improved the sketches after Reclus' corrections, and then Reclus sent them to the editor in Paris.

In 1890 Reclus left Switzerland, having finished the *Nouvelle Géographie Universelle*, and Perron commenced a period of independent work. He compiled maps for a number of French and Swiss geography books published around the turn of the century. He also had to look after a collection of 7000 maps, used for writing the main work, that Reclus had given to him for safe-keeping when he moved to France. In 1904 he was appointed Conservateur du dépôt des cartes de la ville de Genève by the town and on 4 November 1907 he opened a cartographic museum, in the rooms of the Geneva Public and University Library, which seems to have been the first of its kind in the world. He hoped that the institution would serve both international experts and the general public as a centre for information and education. Alas, it was never as popular as Perron had hoped, and the number of visitors gradually declined until the exhibition space was reduced and it was eventually closed in 1937.

During their later years Reclus and Perron had little contact with each other. Interestingly, in 1898 Charles Perron was offered the post of Professor in relief modelling at the Institut Géographique de l'Université Nouvelle de Bruxelles (the New University of Brussels), and his name appeared in the programme for the academic year 1898-9. But it is certain that he never started teaching in Brussels. At that time of his life he could not consider leaving Geneva.

He died of influenza on 7 March 1909, at the age of 72 years, having been ill for only one day.

2. Ideas and Influence

An interesting essay dates from the year 1868, during Perron's radical period. It was entitled *De l'obligation en matière d'instruction*, and shows the way in which he was thinking about compulsory education and the subjects to be taught at school – for the social renewal that he hoped was to come. In this essay he decided that human ignorance was the source of all evil, and the root of war, poverty, social disorder and prejudice. 'Social order', he concluded, 'means the complete education of everybody.' Scientific education for every human being, he argued, would make exploitation of every kind (political, religious and economic) disappear. Education of children should be free, and guaranteed by the state.

Besides his important work of providing the bulk of the cartography for Reclus' *N.G.U.*, and the establishment of the ill-fated cartographic museum, Perron's main working field after his separation from Reclus was the construction of relief models. He was convinced that relief models were the most suitable instruments to convey the configuration of the earth. Relief models do not have inherent difficulties in showing the third dimension of space. His aim, throughout the whole of his life, was to convey accurate information, by, for example, eliminating distortions caused by cartographical projection or error. He saw his role as cartographer and as a constructor of relief models as conveying scientific facts in a popular way.

Initially Perron was concerned particularly with relief models for their own sakes, and in using them to construct maps based on the *photography* of relief models. He called this procedure, presented for the first time to the Geographical Society of Geneva in 1894, *Cartographie à terrain vrai* or *Cartographie reliéviste*. After the turn of the century he referred to his method as *Cartographie nouvelle*.

In the final quarter of the nineteenth century relief models were internationally popular, and indeed in Switzerland a plan was even considered for the modelling of the whole country at a large scale (especially 1:25,000). The project foundered because of the cost. Perron succeeded, on his own initiative, and making substantial personal sacrifices, in realizing a scheme to build a plaster model of the whole of Switzerland at a scale of 1:100,000.

The then new official maps of Switzerland at scales of 1:25,000 and 1:50,000 – the *Topographischer Atlas der Schweiz* – were taken as the basis for the geomorphic form of the model, as they had contours. (These maps were also called 'Siegfried-Karten' after their originator Herman Siegfried, 1819–1879.) It was possible to transfer these contours from the maps to the edges of elevation layers on blocks of plaster, and subsequently cut them out. By using a pantograph Perron could transfer the contour lines and at the same time reduce the scale from 1:25,000 and 1:50,000 down to 1:100,000. The pantograph had a leading-pin on one side and was linked to a small drill on the other to cut out the layers of plaster. The final surface was moulded by hand. The method had been previously used in France. Perron built, between 1897 and 1900, a large relief model of Switzerland at a scale of 1:100,000, without any exaggeration in height, and corrected for the globe's curvature. It was awarded first prize at the Paris World Exhibition in 1900. The original was given to the town of Geneva, and still exists. Other copies of this excellent visual aid, including some with geological colouring, were made, but like all relief models it suffers from the disadvantage of being difficult to handle and expensive to produce. Other models made by Perron included *Relief du pays de Genève* at 1:50,000, built in 1898, and *Relief du Salève*, at 1:25,000, which appears to be missing.

Charles Perron was a self-taught cartographer who was not in close touch with the main centre of cartography and relief modelling in Zurich, and shortly after the start of his most famous project, he was attacked by adherents of that famous school, including Albert Heim (1849–1937), and Fridolin Becker (1854–1922). Support for which he had applied in 1896 from the Swiss Office of Cartography (Topographischer Bureau der Schweiz) was refused, on the instigation of his opponents, by the Swiss Parliament, and the matter developed into a minor political crisis, as Perron's supporters and opponents were divided along ethnic lines – a Swiss German party and a Swiss French party. Deeper political divisions in Swiss life suddenly became evident.

During the last years of his life, Perron's main field of activity was the development of what he called *Cartographie nouvelle*. Perron took photographs of his relief models under artificial oblique light. Maps at scales of 1:250,000 and 1:500,000 were reproduced. He believed that these maps, with their vivid representation of the mountains and valleys and other features of the earth's surface, were unsurpassable, but some argued that they were not true maps, as they did not have an accurate representation of vertical height. Perron saw his photo-maps as giving an impression of the view from space. Possibly they should be regarded as the precursors of modern satellite images: however, these maps, like the very large relief models he created, have been largely forgotten.

Bibliography and Sources

1. WORKS ON CHARLES-EUGÈNE PERRON

Dunbar, G.S., *Elisée Reclus, Historian of Nature*, Hamden, Connecticut, 1978.

Jud, P., *Elisée Reclus und Charles Perron, Schöpfer der 'Nouvelle Géographie Universelle'*, Zurich, 1987.

Nettlau, M., *Elisée Reclus: Anarchist und Gelehrter (1830–1905)*, Berlin, 1928.

Nettlau, M., *Anarchisten und Sozialrevolutionäre: Die historische Entwicklung des Anarchismus in den Jahren 1880–1886 (Geschichte der Anarchie*, Vol. III), Berlin, 1931.

Schweizerisches Künstler-Lexikon, Vol. II (H–R), Frauenfeld, 1908.

2. WRITINGS BY CHARLES-EUGÈNE PERRON

1862 (with Georges Perron) *Types militaires de l'armée suisse*, F. Charnaux, Geneva.

1868 *De l'obligation en matière d'instruction*, Association du sous pour l'affranchissement de la pensée et l'individu.

1900 *Des reliefs en général et du relief au 1 : 100,000 de la Suisse en particulier*, Stapelmohr, Geneva.

1900 *Relief de la Suisse au 1 : 100,000*, Comptoir minéralogique et géologique suisse, Geneva.

1904 'Collection cartographique de la Bibliothèque [de Genève]', *La Globe: journal géographique, organe de la société de géographie de Genève*, Vol. 43, 38–45.

1904 'Reliefs à grande échelle', *La Globe: journal géographique, organe de la société de géographie de Genève*, Vol. 43, 103–10.

1907 *Catalogue descriptif du musée cartographique*, Dépôt des cartes de la ville de Genève, Geneva.

1907 *Une étude cartographique. Les mappemondes*, Editions de la Revue des idées, Paris.

1907 'La cartographie', *Revue des idées*, 4th year (15 May), 400–39.

3. MAPS AND RELIEF MODELS

1876–94 56 full-page and approximately 2700 small maps (inserted in text) for E. Reclus, *Nouvelle Géographie Universelle: La Terre et les hommes*, 19 vols, Hachette, Paris.

1890 Approximately 500 maps and diagrams for F. Schrader, F. Prudent and E. Anthoine, *Atlas de Géographie moderne*, Hachette, Paris.

1891 125 maps, plans and diagrams for W. Rosier, *Géographie générale illustrée*, Vol. 1, *Europe*, Payot, Lausanne.

1897–1900 *Relief de la Suisse*, 1 : 100,000; Museum of Natural History, Geneva.

1898 *Relief du pays de Genève*, 1 : 50,000; Museum of Natural History, Geneva.

1901 *Cartographie nouvelle. La Suisse. Carte muette d'après le relief au 1:100,000 construit selon la courbure terrestre*, 1 : 250,000 (2 sheets), Comptoir minéralogique et géologique suisse, Geneva.

1901 *Cartographie nouvelle. La Suisse. Carte muette d'après le relief au 1:100,000, construit selon la courbure terrestre*, 1 : 500,000 (1 sheet), Comptoir minéralogique et géologique suisse, Geneva.

1908 *Relief du Salève*, 1 : 25,000. Present location unknown.

Chronology

1837 Born 6 December, Petit-Saconnex, Geneva, Switzerland

c. 1857–62   Lived in Russia for five years

1860s   Involvement with radical and revolutionary groups in Geneva

1868   Published *De l'obligation en matière d'instruction*

1874   Commenced collaboration with Elisée Reclus. Over 2700 maps produced for *Nouvelle Géographie Universelle*

1890   Collaboration with Reclus ended

1897–90   Largest relief model completed

1890s   *Cartographie nouvelle* developed

1898   Appointment offered at New University of Brussels; duties never taken up

1909   Died of influenza, after one day's illness, 7 March

Peter Jud is a teacher of geography at a high school near Zurich. He is a member of the managing board of the Geographical Society of Zurich.

Pedro C. Sánchez Granados

1871–1956

Héctor Mendoza Vargas

Pedro C. Sánchez has a significant place in Mexican geography because of his work and teaching in the fields of geographical survey, geodesy, mapping, and also the diffusion of geographical determinism. The period of his life spans two Mexicos: those of before and after the Mexican Revolution (1910–17).

In that historic event early in the twentieth century a new political and economic order was brought into being. It was a difficult time for Mexico, with old administrative procedures being swept away, and it culminated in the bringing forward of a new political constitution for the country in 1917. Geography was a part of this process of change, for there was an urgent need to evaluate the resources of Mexico's extensive land-mass. Pedro C. Sánchez was an enthusiastic participant in these activities, his work in geography including both academic investigations and the adaptation of geographical techniques to the problems of the country. Another important part of his work was his international contribution as organizer and founder of the Instituto Panamericano de Geografía e Historia (Pan-American Institute of Geography and History, PAIGH).

1. Education, Life and Work

Pedro C. Sánchez was born on 19 May 1871, in Durango, Mexico, a region that had had a strong mining tradition since the colonial period. He had five brothers. Being orphaned made his early years difficult. At the age of eleven he was admitted to Instituto Juárez in the state capital: he completed his high school studies in 1886, receiving a gold medal for his academic achievements.

His father had been a miner, and it was perhaps for this reason that he decided to study mining engineering in Mexico City. From 1887 to 1892, therefore, he studied mining and geography at the Colegio de Minería (School of Mines), doing his practical work in the mining district of Pachua, Hidalgo. He commenced teaching topography, mathematics, physics and mineralogy at the Colegio del Estado (State College) in 1892.

In his first mining post Sánchez was given charge of the 'La Blanca' and 'El Chico' mines in Pachuca, Hidalgo, 95 km to the north-east of the capital; this centre had a long tradition of metal-production. Thus from 1892 until 1896 he had extensive experience of underground work in the mines. But at the age of 25, after several years of daily danger and life in an isolated locality, he resolved to return to Mexico City.

After his arrival he soon found a position, because of the support of an old friend from Durango, José G. Aguilera (1857–1941), who was Director of the Institute of Geology. Even though Sánchez was young, he had enough experience in the mines to write, with Aguilera and Ezequiel Ordóñez (1867–1950), a technical monograph about the mining district of Pachuca, giving details of the fracture system, the machinery, mine drainage and metallurgy.

From this point onwards Sánchez abandoned the mining industry. Shortly after moving to Mexico City he joined the National Astronomical Observatory. Here he was able to obtain experience of the geographical and survey work of the Mexican government. The Observatory exchanged telegraph signals with various survey parties from the Comisión Geográfico-Exploradora (Geographical Exploration Commission); this was necessary to fix the geographical co-ordinates of points all over the country.

Later Sánchez worked as chief engineer at the Comisión Hidrográfica del Valle de México (Hydrographic Commission of the Mexico Valley), a post that linked him with the new Secretaría de Comunicaciones y Obras Públicas (Ministry of Communications and Public Works), and was concerned with developing the water supply of Mexico City.

Again this proved to be a short-term appointment, as he was soon designated head of the technical section of the Cadastral Survey of Mexico City (1899). This new office was one of the responses to the rapid urban growth that was taking place at the time: the main objective was to conduct surveys to calculate the areas of land holdings in Mexico City, and the various municipalities that made up the Federal District, so that they could be taxed. The triangulation and polygonation of the city were necessary in order to compile a city map for tax purposes.

As the result of his work, between 1896 and 1902 Sánchez came into contact with a wide range of scientists: these included experts on mining, geology, hydrology, and topographical and cadastral survey: moreover, at the age of 31 he had good experience of precision surveying. With that background and experience he was accepted into the Comisión Geodésica Mexicana (Mexican Geodetic Commission) with Ángel Anguiano, and now began the most important period of his professional life as a geographer.

He had studied geodesy at mining school, and had published several articles on the subject in Mexican scientific journals. Through the work of the Commission, he was able to apply this knowledge on a large scale to the whole of the Mexican land-mass.

At this time, Mexico had concluded an international agreement with the United States and Canada for the measurement of the meridian arc, 45° in length, along the meridian 98° west of Greenwich. This traverse was hundreds of kilometres in length, across the territory of three countries. It represented a project of unprecedented magnitude for the Americas. In essence, the work was divided into several sections, including geodetic triangulation, base measurement, geographical positioning, precision levelling, gravity measurements and the integration of these observations. The actual field observations commenced in 1899, measurement continuing over the next ten years through Oaxaca, Puebla, Hidalgo, Tlaxcala, Queretara, San Luis Potosí and Tamaulipas. Much of this was in mountainous territory, in some cases virtually unexplored, and with severe difficulties of access.

Sánchez was assigned to the fieldwork. In the course of several trips, he came to know a good deal of Mexico's territory. Many times it was necessary to take immediate, on-the-spot decisions in selecting the most suitable mountain peaks upon which to place instruments to obtain the precise values for geodetic triangulation. He also participated in the calculation of geographical co-ordinates from the data obtained.

In 1904 Pedro C. Sánchez was appointed Assistant Director of the Geodetic Commission. In this new position he began studies of gravity at several places in Mexico. These studies proved useful in the detection of oil and mineral resources, as well as in building up a detailed knowledge of the shape of the earth, as part of an international programme sponsored by the International Geodetic and Geophysical Union.

By 1910 the fieldwork of the Commission was almost finished. Measurement had reached 10° 00' 57.27" on the Mexican sector of the 98° meridian. However, the Mexican Revolution started in that year: it was initially a popular movement, but it soon had control of the centre of the country. During the conflict Sánchez was appointed Director of the Geodetic Commission (in 1913); despite difficulties, to prevent later loss of time, he continued to work in Mexico City, revising and correcting the data.

In time, the popular uprising became a bourgeois revolution that not only obtained power, but gave shape and content to a new political constitution (1917). The new government, of course, needed to know as much as possible about the vast Mexican territory. In 1915, at the age of 44, Sánchez was invited to collaborate with the new administration. President Venustiano Carranza asked him to plan and head the Dirección de Estudios Geográficos y Climatológicos (Office of Geographical and Climatological Studies). The new organization had the funding and technical expertise to initiate a new programme of geographical work throughout the country, and was the centre that set the course of Mexican geographical research in the service of the revolutionary government. By 1916 the geodetic network had been completed, along with that on the 98° meridian. With the available funds and resources, Sánchez started on three projects: a new Geographical Atlas of Mexico, a general map of Mexico, and a series of thematic maps of the country.

The Geographical Atlas (1919-21) provided a new vision of Mexico through 32 maps: one of each state, and two general maps of the country, at scales of 1:2,000,000 and 1:5,000,000 respectively. These included much new information about the topography and hydrography of Mexican territory. The Atlas was intended for those concerned with public administration, especially those concerned with hydrology, irrigation and agriculture. The first edition remained in print until 1942, being reprinted 24 times.

A map of Mexico City was published in 1918, as were others in the years that followed; maps of Durango (1919), Sonora (1924), Hidalgo (1926), Mexico state (1927), Chiapas (1927), Chihuahua (1927) and Aguascalientes (1934); also other maps of Jalisco, Hidalgo, Morelos, Nuevo León, Veracruz, Tlaxcala and San Luis Potosí. Thematic maps at a scale of 1:5,000,000 (geology, hydrology, climate, magnetism, municipal divisions, political divisions and communications) were also published completing the portrayal of the essential data of Mexican geography.

In order to complete detailed coverage of the country, Sánchez planned to produce a series of maps of Mexico at a scale of 1:500,000 in 50 sheets. This was effected between 1924 and 1942 by the Dirección de Estudios Geográficos y Climatológicos, each sheet being 2° of latitude by 3° of longitude. At the same time new information was being published in a series of technical reports on the geodetics, topography, climatology, hydrology, gravity measurements and geography of Mexico.

Sánchez personally informed international conferences of these achievements. He therefore visited Europe for the meetings of the International Geodetic and Geophysical Union in Rome (1924), Prague (1927 and 1930) and Stockholm (1933); on each of these occasions he distributed a printed summary of the methods and results of these projects.

As regards cartographic design he was influenced by the writings of L. Defossez and Arthur R. Hinks, but because at the time there were no books appropriate to the needs of Mexico, he published in Spanish *Apuntes sobre Cartografía* (*Notes on Cartography*) (1928). This work discussed the theory of maps as applied to the shape and dimensions of Mexico's land-mass, along with the mathematics of map projections.

At the age of 57, Sánchez was only too aware of the continuing importance of geographical survey in Mexico. Besides the government-imposed funding difficulties, there was the natural difficulty resulting from the size of Mexico's territory (2 million km^2). The problems were similar in many of the other countries of the American continent.

Even the USA, at the end of the First World War, had no adequate topographical maps of much of its territory, particularly along the border with Mexico. From Argentina to Canada, maps were uneven in coverage, and of uneven quality. What maps existed were the result of the activities of each government working in isolation. Between 1920 and 1945, however, the American Geographical Society completed the *Map of Hispanic America*, in 107 sheets, at a scale of 1 : 1,000,000. The co-operation between the Society and the various governments that this involved was most successful.

At the Sixth Pan-American Conference (Havana, 1928) Pedro C. Sánchez proposed a motion seeking the establishment of a Pan-American Institute of Geography, to co-ordinate geographical research and mapping among the American states. The Cuban delegate suggested that history be integrated into the new organization. Mexico City was suggested for the location of the permanent headquarters of the Instituto Panamericano de Geografía e Historia (the Pan-American Institute of Geography and History, PAIGH): with the attendance of ambassadors and dignitaries from many countries, the building was opened by the Mexican President, Pascual Ortiz Rubio. The two-storey building in Tacubaya, with a central patio, had ample meeting rooms, offices and a library.

Despite this optimism, difficulties developed. The delegate from Cuba considered that PAIGH should be at the service of all nations in the Americas, and should assist in the cases of border disputes. Argentina and Venezuela postponed their participation. Only Brazil, Colombia, Costa Rica, Cuba, Chile, Ecuador, El Salvador, Guatemala, Honduras, Mexico, Nicaragua and Peru were initially associated.

But the Institute also secured the participation of William Bowie (1872–1940) as first President: he was an expert from the US Coast and Geodetic Survey. Sánchez was the Financial and Administration Director of the new centre, in which there were two geographical sections: (1) topography, geodesy, cartography and geomorphology, and (2) human geography, ethnography, historical geography, biogeography and economic geography. History similarly had two subdivisions: (1) prehistory, pre-Columbian history, archaeology and colonial history, and (2) post-independence history in the Americas.

In 1934 Pedro C. Sánchez left his position in the Dirección de Estudios Geográficos y Climatológicos. At 63 he had been responsible for the geographical office of the Mexican government for over twenty years, and his influence was amply recognized. He then commenced work on a full-time basis in PAIGH, moving from a national perspective to a Pan-American view of geographical problems. He had to establish the financial basis of PAIGH in order to ensure its permanence: in this his political contacts were important.

At the meeting in Brazil in 1932 monetary fees were set for each country, and it was proposed that the Second General Assembly of PAIGH be held in Washington, DC, in October 1932. The active participation of the USA was announced. Wallace W. Atwood, the new President of PAIGH, gave his opinion on the evolution of, and innovation in, geography in those days:

> In the field of geography, a very ancient science, but one that has expanded during the last few decades to include the study of man's adjustment to his environment and the influence of the environment upon the occupations and prosperity of nations, there are wonderful opportunities for cooperation. Modern geographers are deeply interested in the problems of the conservation and the wise utilization of lands. Geographers are giving special attention to the distribution of population ... the amount and distribution of fuels, of all mineral reserves, of water power, forests, pasture lands, farm lands and of all forms of plant and animal life on land or in the sea, fall into the field of geographical research. Cooperation in the preparation of maps and charts and in the gathering of climatic data are included in the program of the Pan-American Institute of Geography and History. (Atwood, 1935, p. 117)

The participation of the United States was important, not only because of the financial support provided, but because of the contribution of various experts from that country, particularly in the fields of geodesy and mapping.

PAIGH promoted a number of other projects such as the organization of the Pan-American Library of Geography and History, the publication of the *Geographical and Historical Yearbook of the Americas*, the compilation of a geographical bibliography of the Americas and the mapping programme. These activities were organized with the active participation of delegates from each national committee, securing the highest degree of co-operation among the states of the Americas.

Pedro C. Sánchez was particularly concerned with the mapping of the Americas project, and promoted the proposal with enthusiasm at the general assemblies of PAIGH in Rio de Janeiro (1932), Washington (1935) and Lima (1941). An important aspect of this proposal was the securing of adequate triangulation in Central America, thereby connecting the geodetic networks of North and South America. This plan was also included in the objectives of the International Geodetic and Geophysical Union, in which Bowie also participated.

The map of the Americas was important, in the years after the Second World War, in planning economic development programmes. Linked to this was the proposal by Preston E. James, the very active PAIGH representative of the United States, for a series of surveys for the Americas, at a variety of scales, based on the region concept, and emphasizing the study of landforms, climate, population and use of land. The co-operation of geographers from the many countries involved was necessary. Preston E. James described his plan:

> The objective of the survey is to provide a base of organized geographical knowledge regarding the nature of the land and its resources and the existing relationship of the people to the land. This involves the selection of relevant categories of information regarding the surface, the soil, the water, the

climatic conditions, the vegetation surface, the actual and potential resources, the density and patterns of population, and the nature of the present use of the land. (James, 1952, pp. 75–6)

The plan was adapted somewhat, and integrated with other initiatives, and became Project 29 of the Organization of American States (OAS), Technical Cooperation on Natural Resources of the Americas. Robert R. Randall and Pedro C. Sánchez directed the project. Between 1952 and 1953, a group of experts headed by Jorge A. Vivó (1906–1976) completed investigations on geodesy, geology, cartography, meteorology, soil and vegetation. The first results, on Mexico and Central America, were published in a series of five books and maps in 1954 and 1955.

One of Sánchez's tasks was the diffusion of knowledge about the projects and the results of the investigations. For this, PAIGH's publication *Revista Geográfica* was brought into being; it was able to bring into contact with one another the dispersed geographical community of the Americas. The first edition was published in 1941, partly through the efforts of the Spanish scientist Pedro Carrasco, with the assistance of Manuel Medina and Jorge A. Vivó, both of the National University of Mexico. *Revista Geográfica* became an important journal with a wide circulation.

Another important project promoted by Sánchez was a geographical bibliography of each country: some of the results of this enterprise were published in *Revista Geográfica*, but in the case of Mexico, the book *Bibliografía Geográfica de México* was published in 1955: this comprised a register of 4600 items – atlases, books, brochures and articles on the geography of Mexico – arranged 'from the general to the particular: from the great zone to the minor region, from a state to an important zone in that state' (Bassols Battala, 1956, p. 161).

The library was another initiative of PAIGH, and Sánchez was active in its formation. Books, journals and maps were received from many institutions in the Americas and Europe. The first Director was Jorge A. Vivó, a young Cuban who arrived in Mexico in 1938; the library expanded rapidly.

The teaching of geography also had a special place on PAIGH's agenda. At a meeting in Washington, DC, in 1952 a report was presented on the academic position of geography. While in the United States and Canada the situation was excellent, and there was 'wide practical use of geography' in those countries, the situation in Argentina, Brazil and Mexico was poorer, with limited use being made of the subject. In a third group of countries – Chile, Ecuador, Panama and the Dominican Republic – there was a general lack of organization and very limited use made of geography by governments in administration and urban planning. Faced with this very uneven picture, PAIGH became active in the promotion of geographical studies and research throughout the Americas. Sánchez vigorously promoted the subject at the National University of Mexico.

By 1955 PAIGH had completed the first 25 years of its life, a period sometimes known as the 'experimental' phase of organization and consensus-building. The support of governments had been indispensable. From his position as Director, Sánchez had displayed his talents and wide influence in the promotion of its many projects. At the age of 83, however, in 1954, he retired from the position of Director, and was appointed Counsellor and Adviser. He died in Mexico City on 17 March 1956.

2. Scientific Ideas and Geographical Thought

Pedro C. Sánchez completed courses to qualify him as a 'geographical engineer' at the Colegio de Minería (Mining School), and then successfully applied the techniques in the Mexican Geodetic Commission. This work was complemented by teaching (engineering and physical subjects, 1905–14) at the Escuela Nacional de Ingenieros (National School of Engineering) and geodesy at the Colegio Militar (Military College) in 1912 and 1913.

The revolutionary government, around 1915, encouraged the teaching of geography at the National University of Mexico. Sánchez started teaching this subject, from a completely different point of view from his previous 'geographical engineering' experience. He was quite perceptive of new ideas in geography, particularly from Europe, e.g. those of Friedrich Ratzel (1844–1904; *Geographers*, Volume 11), Ferdinand von Richthofen (1833–1905; Volume 7), and above all Paul Vidal de la Blache (1845–1918; Volume 12) and his pupil Jean Brunhes (1869–1930) on the subject of human geography and Emmanuel de Martonne (1873–1955; Volume 12) on physical geography (Martin and James, 1993, pp. 189–202). The influence of Charles Darwin (1809–1883; Volume 9) can also be detected. Thus Sánchez became a receiver and diffuser of ideas from European geography into Mexico. Geography was, for him, 'the science which studies the distribution in space of, and the interactions between, the physical phenomena on the terrestrial surface'.

All these influences were integrated into Sánchez's geographical vision. In his book *Geografía Física* (Physical Geography) (1927) he incorporated the dynamic concepts and innovations concerning landforms, such as the notions of William Morris Davis (1850–1934; Volume 5), as well as some of the ideas of Emmanuel de Martonne. In the first part of the book he presents a systematic study of erosion (an external agent) and tectonics (an internal agent) as factors in the development of landforms. In chapter 7 he applies these concepts to the classification and evolution of Mexican landforms, incorporating also the concept of the hydrological basin. In the latter part of the book, the influence of geology in his work is clearly seen, in the treatment of geological time, palaeogeography, the geology of Mexico and the classification of mountains.

About 1930, an interest in geophysics becomes evident in his work – in particular the vulcanism and seismology of the Mexican Republic, although this did not imply his abandonment of geography, as *Geografía humana* (Human Geography) was published in 1930. In the search for general laws that rule the relations between humans and the earth, and the emphasis on the distribution of population on the surface of the earth, the influence of Vidal de la Blache and Brunhes can be detected, but also that of Ratzel, from whom he inherited ideas of geographical determinism.

He also presented these ideas in his classes, especially those of geographical determinism and positivism. From time to time he gave free public lectures where he expounded his somewhat controversial views on the influence of geographical factors on the development of nations. At the age of 75 he wrote:

> Because of the hot climate, races in the tropics are lazy and progress very little. The same happens in cold climates; peoples from northern Asia are almost stupid and not susceptible to progress. (Sánchez, 1946)

To these direct geographical factors he added the indirect influences of diseases such as malaria and yellow fever. These influences, in his view, gave a full explanation of why there would always be powerful and weak, rich and poor, and of how some 'populations enforce dominion over their weaker and less active neighbours'. This provided, to him, a justification for war as a necessary element in slow human evolution. At the Third Mexican National Geographical Congress in Guadalajara in 1942, he presented a paper explaining the invasion of Mexico by the United States (1847–8) as a 'consequence of potentiality' of the northern neighbour, and the loss of territory in terms of the 'limited resistance' of Mexico. In similar vein he went on to expound the view that the Second World War established 'a new era for humanity' in which the United States, triumphant and rich, would become the determining force and 'luminous guiding light' carrying humanity along the paths of progress.

3. Influence and Spread of Ideas

The most important influence of Pedro C. Sánchez was accomplished through his flair for the organization and administration of geographical work, first in Mexico and later throughout the Americas through PAIGH. As the result of his direct contact with the Mexican government authorities, he was without doubt the most influential geographer in his country at the time. He was closely involved with the main geographical projects of both Mexico and the entire American continent.

He was also important in the formation of professional geographers as a group. Under his guidance, the Department of Geography of the National University of Mexico was created in 1936: he gathered around him a group of former geographical engineers and other scientists, including exiles from Spain, to form the founding group of the profession, which took over the design and organization of courses, including general geography, human geography, biogeography, the geography of Mexico, geology, geodesy, meteorology, statistics and cartography (Vivó and Riquelme, 1961, pp. 11–54). Geographical teaching was profoundly influenced by the natural sciences, and was taught within the Faculty of Science at the National University from 1939 to 1943, in which year geography returned to the Faculty of Philosophy. As the result of the influence of Sánchez, between 1941 and 1955 71 students obtained their diploma, 25 took master's degrees, and four obtained a doctorate in geography.

Undoubtedly Pedro C. Sánchez was able to adapt geography to the public administration of Mexico, and this work gave him substantial influence with the government. In spite of this, his deterministic academic views were widely expressed in his published work, at conferences, in public meetings and in his teaching of human geography at the National University of Mexico.

By 1950, however, there emerged another generation of geographers, self-named the 'transition' group, who did not sympathize with these extreme deterministic views. A very different geographical vision, that of possibilism, opened up a new era for the study of geography in Mexico.

Bibliography and Sources

1. OBITUARIES AND REFERENCES ON PEDRO C. SÁNCHEZ

Martínez Becerril, C., 'Semblanza científica del Ing. Pedro C. Sánchez', *Boletín de la Sociedad Mexicana de Geografía y Estadística*, Vol. 89 (1960), 105–42.

Pedrosa, C., 'Pedro C. Sánchez', *Revista Geográfica*, Instituto Panamericano de Geografía e Historia, Rio de Janeiro, Vol. 18, No. 44 (1956), 115–16.

Vivó Escoto, J.A., 'Pedro C. Sánchez Granados', *Los estudios sobre Recursos Naturales en las Américas. Datos biográficos*, Vol. 4, Instituto panamericano de Geografía e Historia/Proyecto 29 del Programa de Cooperación Técnica de la OEA, Mexico City (1954), 222–3.

Vivó Escoto, J.A., 'Bibliografía de Pedro C. Sánchez', *Boletín de la Sociedad Mexicana de Geografía y Estadística*, Vol. 89 (1960), 143–56.

2. SELECTED PUBLICATIONS BY PEDRO C. SÁNCHEZ

Books

1912   *Técnia de la Geodesia, para uso de los alumnos del Colegio Militar*, Talleres de Departamento de Estado Mayor, Secretaría de Guerra y Marina, Mexico City.

1920   *Cálculo de las probabilidades y teoría de los errores*, Dirección de Estudios Geográficos y Climatológicos, Publicación No. 2, Mexico City.

1927   *Geografía física con aplicaciones a la República Mexicana*, Dirección de Estudios Geográficos y Climatológicos, Publicación No. 7, Mexico City.

1928   (with Bustamente Octavico) *Apuntes sobre Cartografía*, Dirección de Estudios Geográficos y Climatológicos, Publicación No. 29, Mexico City.

1930   *Enseñanzas fundamentales de la Geografía humana*, Dirección de Estudios Geográficos y Climatológicos, Publicación No. 22, Mexico City.

1933   *Catálogo de datos numéricos geográficos y topográficos de la República Mexicana*, Dirección de Estudios Geográficos y Climatológicos, Publicación No. 8, Mexico City. Second edition.

Pamphlets

1928   *Estudio Hidrológico de la República Mexicana*, Dirección de Estudios Geográficos y Climatológicos, Publicación No. 17, Mexico City.

1931   *Geografía Económica*, Dirección de Estudios Geográficos y Climatológicos, Publicación No. 28, Mexico City.

1932   *Geografía Política*, Dirección de Estudios Geográficos y Climatológicos, Publicación No. 26, Mexico City.

1932   *Volcanismo*, Instituto Panamericano de Geografía e Historia, No. 4, Mexico City.

1935   *La evolución de la Geografía*, Instituto Panamericano de Geografía e Historia, No. 12, Mexico City.

1944   *El Instituto Panamericano de Geografía e Historia: su creación, utilidad, desarrollo e importancia de sus trabajos para la América*, Instituto Panamericano de Geografía e Historia, No. 72, Mexico City.

Major Articles

1926   'Proyecto de ejecución de una Carta de la República en 50 hojas a la escala de 1:500,000 [1924]', in Jesús Galindo y Villa, *Geografía de la*

República Mexicana, Vol. 1, Sociedad de Edición y Librería Franco Americana, Mexico City, 416–19.

1928 'The history of geodesy in Mexico', *Transactions of the American Geophysical Union*, Vol. 9, 20–31.

Maps
1919–21 *Atlas Geográfico de la República Mexicana*, Secretaría de Agricultura y Fomento, Dirección de Estudios Geográficos y Climatológicos, Mexico City. First edition. 32 maps plus one general map.

3. OTHER REFERENCES MENTIONED IN THE TEXT

Atwood, W.W., 'The opening of the Second Assembly of the Pan American Institute of Geography and History (Washington, DC, 1935)', *Segunda Asamblea*, Instituto Panamericano de Geografía e Historia, Publicación No. 22 (1937).

Bassols Battala, A., 'Suggestions for a bibliographical classification of geographical interest', *Revista Geográfica*, Instituto Panamericano de Geografía e Historia, Vol. 19, No. 45 (1956), 161.

James, P.E., 'A program for an exploratory survey of the Americas', *Revista Geográfica*, Instituto Panamericano de Geografía e Historia, Vols 9–10, Nos 25–30 (1952), 61–79.

Martin, G.J. and James, P.E., *All Possible Worlds. A History of Geographical Ideas*, Wiley, New York, 1993. Third edition.

Chronology

1871	Born 19 May, Durango, Mexico
1886–92	Moved to Mexico City; studied at School of Mines
1892–6	Worked in silver mines, Pachuca, Hidalgo
1897	Returned to Mexico City; temporary job at Institute of Geology
1897–1902	Worked in the National Astronomical Observatory; later worked on the cadastral survey of Mexico City
1902	Commenced work for the Mexican Geodetic Commission. Measured the 98°W meridian in collaboration with surveyors from Canada and the USA
1904–10	Assistant Director of Mexican Geodetic Commission. Worked on border with USA
1912	Appointed Professor of Geodesy, Military College. Mexican Revolution
1913	Appointed Director, Office of Geographical and Climatological Studies
1918	Appointed Professor of Geography, University of Mexico
1919–21	*New Geographic Atlas of Mexico*; maps of the country at 1:2,000,000 and 1:5,000,000

1922	Visited Europe
1924	Series commenced of maps of Mexico, at a scale of 1:500,000 in 50 sheets
1927	Physical geography text published. Influence of William Morris Davis
1929–30	Foundation of PAIGH
1930	Human geography text published. Influence of Ratzel, Vidal de la Blache and Brunhes
1941	Retired from teaching geography
1955	25th Anniversary of foundation of PAIGH: presentation of gold medal
1956	Died 17 March, Mexico City

Héctor Mendoza Vargas recently completed a doctoral thesis at the University of Barcelona. He is now at the Institute of Geography, National University of Mexico.

Zheng He

1371–1433

Lu Zi and Liu Yan

'Eunuch San Bao's Voyage to the Western Ocean' is a story known to every Chinese family. San Bao, otherwise known as Zheng He, holds a unique position in Chinese geography. Half a century before the discovery of South-East Asia and the lands of the Indian Ocean by Europeans, this great Chinese navigator and explorer commanded a large group of ships to South-East Asia and into the Indian Ocean – the 'Western Ocean' in Chinese terms. He ventured there seven times over a thirty-year period.

Zheng He's exploration of the Western Ocean was a significant event in the Ming–Qing Dynasty. It revealed navigational and ship-building technology in China to be on an equal footing with those of almost anywhere else in the world.

During the first thirty years of the fifteenth century Zheng He's exploration activities extended over nearly one quarter of the circumference of the earth – from 10°S to 32°N, and from 40° to 125°E. Among the most westerly places that he visited was Mogadishu in what is now Somalia. The voyages represented the most ambitious of ancient China's explorations towards the West. Zheng He's achievements expanded the vision of the Chinese people.

1. Education, Life and Work

In the *Ming Dynasty History* Zheng He's life story is briefly covered, as follows:

> Zheng He was born in Yunnan province, and was originally known as Eunuch San Bao. Zheng He worked in the army of Yan King in his early years. Later he joined the rebel forces. He became Head of the Eunuchs.

Born in a remote part of Yunnan province in 1371, Zheng He was raised in Kun Yang county. His real surname was Ma, although he is better known by his given surname, Zheng. San Bao was a nickname, meaning 'the third baby'. His family had been Muslims for many generations, and his father had made the long pilgrimage to Mecca, a fact that was deeply imprinted on Zheng He's mind as a child. Zheng He also was a devout Muslim: records of two important events exist. Before going to the Western Ocean, Zheng He burnt incense on board ship on 16 May 1417, praying for God's blessings. In the records of a mosque in Yangtse Street, Xian city, it is recorded:

> When Eunuch Zheng He received the order to go to the Western Ocean, the translator who went with him was from this mosque: his name was Hassan.

At the time Zheng He was born, the rule of the Yuan Dynasty had been overthrown elsewhere, but the dynasty still ruled in the remote province of Yunnan. Zheng's family suffered badly in the turmoil and chaos of war, particularly after the death of Zheng He's father (when Zheng He was twelve years of age). On a memorial tablet in a tomb, the inscription to Zheng He's father was:

> This person is a Harchi [a member of a particular Muslim sect]. His surname is Ma, and he was born in Yunnan province. He had two sons: the first was named Ming, the second He. He was naturally intelligent. He later took the name Zheng, and was appointed Head of the Eunuchs. In his service of Yan King, he was alert and resourceful, modest and prudent, and thought little of hardships. His talents were widely appreciated. (Yuan Jiaqu, *Yunnan Annals*, Vol. 3)

Zheng He became a eunuch of the Imperial Palace at the age of fourteen. Later the Emperor Yongle changed his name to Zheng, partly because of the Chinese proverb that declared 'Ma [horse] cannot enter the palace'.

From August 1399 to June 1402, a war between Ming Dynasty rulers took place: the Jing Kang war, between Yan King and his nephew, the Emperor. Zheng He distinguished himself in this conflict, and established his reputation as a courageous officer. Yan King eventually established himself as Emperor Yongle (Yung Lo), and appreciating Zheng He's ability, dispatched him as an envoy into the Western Ocean in 1405. Wang Jinghong was appointed as second-in-command. They were given charge of a large group of ships.

Three reasons are given for Zheng He's expedition into the Western (i.e., Indian) Ocean:

1. Some said that the former Emperor, Hui, had been killed in a fire, but the new Emperor Yongle, having taken power by force, believed that Hui might have escaped abroad, and could return. Thus he charged Zheng He with searching for him, as the *Ming Dynasty History* puts it, 'to wipe out the remnants of the enemy'.
2. Mongolia attempted to block the continental trade route from China to the west, and as Chinese industry and commerce sought a large market, alternative routes were sought. The Emperor was also eager to bring 'rare and valuable treasures' to his capital. Other envoys were sent to the north and east.
3. Zheng He was also keen to display Chinese military strength, prestige, wealth and prosperity to the countries of the Indian subcontinent and South-East Asia with whom he came into contact.

Since the Song Dynasty (960–1279), groups of Chinese ships had been sailing along the shore of the South China Sea and parts of the Indian Ocean, and had built up a high level of technical expertise in terms of shipbuilding and navigation. However, it was only in the Ming Dynasty (1368–1644) that Chinese society and the Chinese economy and its financial capacity were strong enough to organize and support very substantial, prolonged, distant voyages. At the same time the Emperors began to allow foreign traders to enter China, as well as to encourage China's trade abroad.

Zheng He was considered to be suitable to lead the major voyage in many ways. He was said to be both a Buddhist and a Muslim; he was familiar with the customs of the Arabian states, and had a good command of Arabic; he was said to be 'wise and to possess exceptional talent'; he had a knowledge of strategy, and of imperial palace politics. He was a strong supporter of the Emperor Yongle, sharing his vision of establishing China as a powerful, prosperous trading nation. For his own part, Zheng He wished to make a pilgrimage to Mecca.

Zheng He's expeditions were enormous undertakings; at their greatest they included 60 large ships and 100 smaller vessels. The largest vessel was described as being 44 zhang long and 18 wide (a zhang was a Chinese metre), and said to be capable of carrying a thousand people. It had nine masts and twelve sails. There were said to be 27,000 individuals in the group as a whole, including diplomats, soldiers, traders, priests, technical personnel, translators and medical doctors. The organization was very tight. The fleet was an impressive sight. Zheng He himself described the way in which despite the 'large, hill-like waves of a rough sea' the ships 'raised high masts ... and cleaved through the raging waves ... like someone taking a stroll along a road'.

The first voyage began in July 1405. The group of 62 large ships, loaded with silk, chinaware, tea and gold, departed from Liujia Bay in Jiangso province. They passed through the Taiwan Strait and the South China Sea and entered the Indian Ocean through the Malacca Straits.

The second voyage was undertaken to bring home some of the foreign envoys. He took advantage of the north-east monsoon, and sailed to Java, Sri Lanka and Thailand.

The third voyage included only 45 ships. It reached what are now India and Pakistan. It was on this trip that Zheng established a town at Malacca, with enclosing walls, gates and four drum-towers. This became a base for expeditions on his later trips. On this voyage Zheng He had a confrontation with the King of Sri Lanka, and eventually captured him and his wife and transported him to China under escort.

The fourth voyage again visited India, but included some places not previously visited, including Hormuz and the Persian Gulf, as well as places in Sumatra and Java. The *Ming Dynasty History* records that in Sumatra Zheng He became involved in local squabbles: the former ruler Suganci planned to take power, and became angry with Zheng when he would not support his plans. Zheng's ships were attacked, but Zheng struck back and captured Suganci.

On the fifth and sixth voyages, Zheng's flotilla went further than ever before, and arrived on the east coast of Africa, south of the Equator, bringing back lions, leopards, giraffes and ostriches.

Since childhood, Zheng He had wanted to make a pilgrimage to the Muslim holy city of Mecca; he had been persuaded to make this journey by his father. On his seventh voyage, Zheng He had his wish, reaching Mecca via the Red Sea. He visited the mosque, shed tears and shouted 'Allah is great!' He kissed the holy stone. During this voyage, a group of ships branched off from the main flotilla and visited Persia. Alas, Zheng He died in the course of the return leg of this voyage, probably in India, and was buried at the foot of a mountain; although there is also a site in

Java, said to be his burial place, that has itself become something of a centre for pilgrimage.

2. Scientific Ideas and Geographical Thought

Although Zheng He's voyages and explorations were of momentous importance, his influence extending over an area that included 37 modern countries, they have to be seen in the context of a long history of the development of geographical knowledge in China, which has some writings coming from remote antiquity – the Zhou Dynasty (*c*. 1100–771 BC).

Zheng's important contributions were in marine geography and in enriching China's knowledge of the coast, in surveying, and in marine meteorology. Perhaps his major contribution was his series of nautical charts. These were published in a book with a 24-page preface, twenty pages of charts and two pages of details of the stars. The charts include some 500 place names, from the China coast to that of East Africa. Details of the coast, islands, sandbars, shallows, coral reefs and bays are shown. The form of the map is unconventional to the modern eye, not being arranged in the 'north at the top, south at the bottom' format. It is in elongated form, relationships being shown by bearings. Thus the first nine maps show the Chinese coast. The eleventh shows Cambodia, the twelfth Java and Thailand, the fourteenth shows Malacca, the eighteenth Sri Lanka and parts of India; Aden is shown on the nineteenth, Hormuz on the twentieth. Many of the islands of the South China Sea are shown in detail.

The depth of the sea and the nature of the seabed are shown. Zheng He's sailors attached a hammer to the end of a rope, covered the hammer with grease, and lowered it into the sea; sand and other material from the bed stuck to the hammer and could be examined. The position of the ships was sometimes determined from the depth of the sea and the quality of the seabed.

Zheng He observed weather phenomena including clouds, fog and rainbows; he also used these and the nature of the waves and of marine life to predict storms.

He also enormously extended trade between China and the lands bordering the Indian Ocean. Manhuan, Feixin and Gong Zhen recorded what they saw and heard on the long voyages with Zheng, and wrote three books: *Go Sight-seeing*, *Stars* and *Customs in Foreign Countries*. The books are a careful record of the politics, economy, military affairs, culture, history and geography of every country visited. Climate (including precipitation), soils, distances travelled by sea, hydrology, living things, strange animals, historical sites, legends and stories, cities and towns are mentioned. Here is a paraphrased extract from *Customs in Foreign Countries*:

> The south of the Arabian Peninsula stands facing the sea, backing on to the mountains: the sea is to the south-east, the mountains to the north-west. It produces, especially, frankincense as a resin. Here the climate is like August or September [in China] all the time – not too hot, and not too cold. Rice, wheat, beans and millet are grown, along with all kinds of vegetables, melons and egg-plant. Animals include cattle, horses, donkeys, cats, dogs, cocks and ducks. There are ostriches in the hills. There are camels – both the one-humped and two-humped forms. All the local people ride them and eat them.

The book called *Stars* describes Africa in lively detail. It was described as being 'arid, with little rain' and the residents as having curly hair and 'building houses with stone'. Animals such as elephant, lions and leopards are mentioned. Zheng He's expedition was thus exploring and recording Africa half a century ahead of anyone from Europe.

3. Influence and Spread of Ideas

Zheng He's seven voyages were thus major events in the navigational history of the world. They reveal a level of seamanship of a high order. He opened a new route from East Asia towards the West and Arabia, and was the first person from China to reach East Africa. Nevertheless his direct influence on geographical thought was not very strong. His method of drawing charts, often at small scale, without latitude or longitude, was not satisfactory. Also, it was still believed that the world was flat, and thus Zheng's maps and discoveries are without context in terms of the total world map.

But the voyages had a very important influence on the development of China's trade and communications, contributing to the expansion of manufacturing. No longer was China dependent for its trade with the West on the continental route or 'Silk Road'. The chinaware and silk industries, in particular, expanded massively. This export orientation was connected to the expansion of the capitalist mode of production in China.

Also, although there had already been some movement of Chinese people into South-East Asia before Zheng's day, Chinese settlement expanded as the result of Zheng's voyages.

On the negative side, Zheng's voyages were enormously expensive, particularly as the ships were laden with gold and silver and other treasures for Zheng to present as gifts to the rulers of the countries he visited. When a new Emperor was enthroned in 1424, he ended the 'treasure ships policy' of sending massive expeditions to the Western (Indian) Ocean. It was described at that time as a 'vile and evil policy that impoverished the people and drained the treasury'.

Place names commemorating Zheng He, or San Bao, exist in several parts of South-East Asia; these include a ridge and a temple in Java, a tower in Thailand, and a hill near Malacca.

Bibliography and Sources

1. BIOGRAPHICAL SOURCES

The following contain biographical information on Zheng He:

Deng Shoulin, *Dictionary of Geography*, Hebei Education Press, 1984.

Ding Qian, *The Geography of Foreign Countries*, 1915.

The Emperor Taizong Record of the Ming Dynasty.

The Emperor Renzong Record of the Ming Dynasty.

The Emperor Xuanong Record of the Ming Dynasty.

Hu Shijian, *Chinese Tourism History*, Tourism Education Press, 1989.

Huang Shenghui, *Tributes List from the West Ocean*, 1808.
Jian Fan, *Investigations of Translations*, 1695.
Mao Yuanyi, *Wubei Annals*.
Ming Dynasty History.
Rockhill, J., *Zheng He's Sail*, 1915.
Shang Rong, *Chinese Historical Outline*, People's Press, 1954.
Si Tushangji, *Geography and History of China*, Guangdong Map Press, 1993.
Wang Da Zhou, *Island Annals*.
Yuan Jiaqu, *Yunnan Annals*, Vol. 3.
Zheng Dai, *Collected Biographies of Near Neighbours*, Vol. 214.
Zheng He's Family Tree.
Zheng Xiao, *Events and Investigations of Near Neighbours*, Vol. 2.
Zhu Guozhen, *Emperor Events*.

2. WORKS WRITTEN BY ZHENG HE AND HIS COMPANIONS
Feixin, *Stars*.
Gong Zhen, *Customs in Foreign Countries*.
Zheng He, *Nautical Charts*.

Chronology

1371	Born in Yunnan province
1381	Taken prisoner of war
1382	Father died
1390	Became eunuch
1398	Became Head of the Palace Eunuchs
1399	Armed rebellion
1402	Emperor Yongle (Yung Lo) seizes throne
1404	Named Zheng He by Emperor
1405	Start of first voyage to the Western (Indian) Ocean
1406	Visits to Vietnam, Java, Sumatra, South India
1407	Returned to China (October); commenced second voyage
1408	Returned to China (July); commenced third voyage
1410	Confrontation with King of Sri Lanka
1411	Returned to China (July)

1413	Start of fourth voyage
1415	Returned to China (August)
1417	Start of fifth voyage (May)
1418	Exploration of East African coast, Arabia
1419	Returned to China (September)
1420	Ming capital moved to Beijing
1421	Start of sixth voyage
1422	Returned to China (September)
1424	Emperor Yongle died
1425	Western Ocean policy abandoned
1431	Start of seventh voyage (January)
1432	Visited Mecca
1433	Died, probably in India

Both Lu Zi and Liu Yan are professors in the Department of Geography, Hebei Teachers' University, Shijiazhuang, Hebei, People's Republic of China. Lu Zi specializes in the geography of communications and Liu Yan in cultural geography.

Index

The index is divided into two parts:

1. A general index, including personal names, organizations, conferences, societies, and geographical concepts, theories and research.

2. A cumulative list of biobibliographies which includes all the geographers listed in volumes 1–20 inclusive.

1. GENERAL INDEX

Adair, John   1–8, 81
Aden   122
Africa   10, 80, 121, 122–3
agricultural studies   45, 49
America   80
American Geographical Society   28, 29, 111
American Philosophical Society   29
animals   63–4, 122, 123; *see also* zoology
Antarctic   15–20
anthropogeography   64–5, 70–1, 72
antiquarianism   3–4, 86, 92, 95
Arabian Peninsula   122
Arctic   16–17, 18, 88
Asia   24–37, 72, 89, 94
Association of American Geographers   28, 33, 65
Association Internationale des Travailleurs   103
astronomy   11, 18–19
atlases   45, 50, 79–82
Aubrey, John   77, 79, 81
Australasia   13–21
Austria   69

Bakunin, Mikhail   103
Banks, Joseph   12–16, 19, 20, 87, 88
Barrett, Robert   28
Barton, Benjamin Smith   97
Becker, Fridolin   105
Berne   87
bibliographies, geographical   113
biogeography   115
birth control   62, 63, 64
Bishko, Julian   35
Bishop, Carl Whiting   29, 34
Bligh, William   16, 17
Blome, Richard   79
Borlase, William   86
Bowen, Emanuel   82

Bowie, William   111, 112
Bowman, Isaiah   28, 29
Britain   1–8, 25–6, 32, 33, 78–82, 85–98
Britannia (Ogilvy)   77–82
Brückner, Eduard   69, 70, 72
Brunhes, Jean   114
Buddhism   44, 46–7, 51–3
Buffon, Comte de   87, 88, 91, 92

Cambodia   122
Cambridge   33, 58
Canada   10–11, 17, 18
Carrasco, Pedro   113
Carruthers, Douglas   27
Carter, George   31
cartography   1–8, 77–84, 102–5, 108–13, 115
catastrophism   92, 93
chart-making   2–6, 122, 123; *see also* surveying
Chiang Kai-shek   29–30, 31, 33
China   24–37, 62, 80, 89, 94, 119–23
chronometers   12, 15
circulated queries   3–4, 77, 81, 94
classical literature   78
classification   114; *see also* taxonomy
clergy   58, 64, 92, 94
Clerke, Charles   12, 16–17
climate   70, 71, 88, 94, 114, 122
coastal surveying   2–6, 10–21, 122–3
communism   30–1
contraception   63
Cook, James   9–23, 88
Cordiner, Charles   90
Coxe, William   88
cross-disciplinary studies   72

dancing   77, 78
Darwin, Charles   20, 57, 64, 65, 114
Davies, Revd Hugh   88–9
Davis, William Morris   28, 114
De Martonne, Emmanuel   114
De Planhol, Xavier   36
Defossez, L.   111
demographic studies   57–65
determinism   114, 115
diseases   60, 62, 114–15
Disraeli, Benjamin   92

earthquakes   80, 93, 114
East India College (Haileybury)   58
Eberhard, Wolfram   37
economic geography   70
economic principles   61

education 43–53, 68–72, 104, 113, 114, 115
engravings 80, 81, 82, 87, 95
environmentalism 57, 62, 64
erosion 114
Essay on Population 57–61
evolution 93, 115

famine 60–1, 62, 65, 71
Farmer, B. H. 33
field observation 94
fisheries 5
folklore 45, 97
food production and pricing 59–60, 61, 65
Formosa (Taiwan) 51
Forster, Georg 15, 19, 20, 88
Forster, Johann Reinhold 15, 19, 20, 88
fossils 86, 92, 93
France 71, 87, 92, 93
frontier studies 34–5, 37
Furneaux, Tobias 15–16

Gaubatz, Piper 37
Gause, G. F. 63
Geneva 87, 102–5
geochronology 92–3
geodesy 109–11, 115
Geographical and Historical Yearbook of the Americas 112
Geographical Journal 27
Geographical Society of Geneva 104
Geography of Human Life (Jinsei chirigaku) 44–53
geology 70, 86, 92–3, 114
geomorphology 70, 71
Germany 71, 87
Gesner, Johann 87
glacial climatology 69, 71
Gladstone, W. E. 64
Godwin, William 58, 60
Gong Zhen 122
Gore, John 12, 16, 17
Graves, Richard 58
gravity 110
Great Atlas (Pitt) 1–2
Great Fire of London 79, 80
Great Wall of China 35, 37
green movement 57, 62
Greenwich Hospital 16
Griffith, Moses 87, 88, 92
Gronovius, L. T. 87
Guggenheim Foundation 28

Hague, The 87
Haileybury (East India College) 58
Haller, Albrecht 87
Halley, Edmund 11
Harvard University 28

Harvard-Yenching Institute 28
Hasekura, Rokuemon Tsunenaga 71
Hawaii 17
Heim, Albert 105
Heimatkunde, see local geography
Herman, Theodore 37
Hinks, Arthur R. 111
Hiroshima Higher Normal College 68, 69–70, 72
history 111, 112
Hobsbawm, Eric 36
Hollar, Wenceslaus 81
homeland studies 45–6, 49–50, 51, 52, 70, 72
Hooke, Robert 4, 79, 81, 82
Hormuz 121, 122
human geography 44–53, 70, 111, 114, 115
Humboldt, Alexander von 20
Hume, David 60
Huntington, Ellsworth 28, 34, 35, 69
Huxley, Thomas H. 26, 34
hydrographical surveying 10–21, 122, 123
hydrology 109, 110, 114, 122

illustrations 79, 80, 87, 88, 92, 95
 cartographic 81, 102–5
imperialism 95
India 62, 88, 89, 94, 95, 121, 122, 123
Indian Ocean 119–23
industrial location 49, 96–7
infant mortality 61, 63
information-gathering 94–5
Inner Asian Frontiers of China 28, 29, 30, 34–5, 37
innovation 36
Institute of Pacific Relations (IPR) 28–9
International Geodetic and Geophysical Union 111, 112
Ireland 62, 65, 78, 81, 87
Isle of Man 88–9
Italian Geographical Society 27

James, Preston E. 112–13
Japan 29, 30, 33, 43–53, 68–72
Java 121, 122, 123
Jefferson, Mark 64–5
Johns Hopkins University 29, 30, 32

King, Gregory 77, 79, 81
Kircher, Athanasius 80
Kirwan, Lawrence 32
Korea 31, 51
Koto, Bunjiro 48
Kropotkin, Pyotr (Peter) 103
Kunimatsu, Hisaya 48–9
kyodo-kai 45–6

Kyoto Imperial University   44, 68, 70

La Gorce, J. O.   35
Lattimore, Eleanor Holgate   27, 32
Lattimore, Owen   24–42
Lausanne   25
Leeds University   32, 33
Leiden   87
Lhuyd, Edward   95
Lightfoot, Revd John   88, 89, 90
linguistic studies   68, 69, 70, 71, 72
Linnaeus, Carl   91, 92
local geography, *see* homeland studies
location, industrial   49, 96–7
London   16, 77, 79, 80, 89
Loten, J. G.   88
lottery funding   77–8, 79

McCarthy, Joseph   31
Maeda, Tamon   46
Makiguchi, Tsunesaburo   43–56
Malacca   121, 122, 123
Malte-Brun, Conrad   20
Malthus, Thomas Robert   57–67
map-making, *see* cartography
Marbut, C. F.   65
marine geography and surveying   2–6, 10–21, 122–3
Martin, Martin   2–3, 5, 6
Martineau, Harriet   64
Marx, Karl   34, 64, 65
Mazell, Peter   95
Mecca   120, 121
Medina, Manuel   113
Meigs, Peveril   31
Merchant Taylors Company   77, 78
meridian arc survey   109–10
Mexico   108–15
Mikesell, Marvin   36
mining   108–9, 114
Mitani, Motohiro   46
Mogadishu   119
Mongolia   27, 28, 30, 33, 35–6
Montesquieu, C.-L.   94
Morgan, William   79, 82

Nakano, Takeshiro   72
Nakanome, Akira   68–76
National Geographic Magazine   28
National Geographic Society   28, 35
natural history   19, 87–98, 122, 123
navigation and seamanship   10–21, 120–3
Netherlands   87
Nettlau, Max   103
New Zealand   13–16, 18, 20
Newfoundland   11
Nichiren sect   44, 46–7, 51–3

Nieuhoff, Jan   80
Nitobe, Inazo   45, 46, 51
nomadism   36, 70
North-West Passage   16–17, 18
Nouvelle Géographie Universelle (N.G.U.)   102, 103

Odauchi, Michitoshi   45, 72
Ogawa, Takuji   48, 68
Ogilby, John   4, 77–84
Organization of American States (OAS)   113
Osaka Foreign Language College   71
Oxford University   86, 87

Pacific Affairs   28–9
Pacific exploration   12–21
Pacific war   29, 30
Page School of International Relations   29, 30, 32
PAIGH, *see* Pan-American Institute of Geography and History
Painter, Sidney   32
Pakistan   121
Paley, William   59
Pallas, P. S.   87, 88, 92
Palmer, William   91
Pan-American Conference (Havana, 1928)   111
Pan-American Institute of Geography and History (PAIGH)   108, 113–15
Pan-American Library of Geography and History   112, 113
Paris   32, 87, 105
Patagonia   93–4
patronage   78, 79–82
Pelliot, Paul   27
Penck, Albrecht   69, 70, 71, 72
Pendleton, Robert   30
Pennant, David   89
Pennant, Thomas   85–101
Penrose, Ernest   30
Perron, Charles-Eugène   102–7
Persian Gulf   121
Pestalozzian principles   46, 49
Petty, William   81
photographic relief maps   104–5
physical geography   70, 114
Pitt, Moses   1–2, 4, 80
place names   20
Plot, Robert   81
Political Economy Club   59
political geography   70
population   57–65, 72, 93, 114
positivism   114
poverty   60, 61, 64
Price, Richard   60

printing, early 78, 79–82, 95

queries, circulated 3–4, 77, 81, 94–5

Rasmussen, D. I. 63–4
Ratzel, Friedrich 72, 114
Ray, John 90, 91
Reclus, Elisée 102, 103, 104
regional geography 69, 70, 72, 96–8
relief models and maps 104–5
research, geographical 43–53, 110
Revista Geográfica 113
ribbon maps 82
Richthofen, Ferdinand von 69, 114
Ritter, Carl 48
Royal Central Asian Society 33
Royal Geographical Society (RGS) 27, 28, 32, 33
Royal Navy 10
royal patronage 77–82
Royal Society 3, 5, 20, 80, 81
  Fellows 2, 16, 59, 86, 87
  Philosophical Transactions 86
Royal Society of Literature 59
Royal Society of Uppsala 91
Russia 30, 92, 102; *see also* USSR

St Bees School, Cumberland 25–6
Sakhalin 70, 72
San Bao, *see* Zheng He
Sánchez Granados, Pedro C. 108–18
Schafer, Edward 37
Schmidt, Wilhelm 69
school geography 43–53, 70
Scotland 1–8, 88, 90, 95
scurvy 10, 12, 14, 16, 19
seismology 80, 93, 114
shells 5
Shiga, Shigetaka 45
Shintoism 47, 51, 53
Sibbald, Robert 2, 4, 5, 6, 81
Siberia 30
Siegfried-Karten 104–5
Silk Route 37, 123
Sinkiang 28
Slezer, John 2, 6
Smith, Adam 59, 60
social order 46–7, 53, 104
Society of Woman Geographers 32
soils 65, 88, 122
Soka Gakkai 44, 46–7, 51–3
Solander, Carl 12, 14, 19, 20
Somalia 119
Spengler, Oswald 34, 35, 36
Sri Lanka 121, 122
Statistical Society 59
statistics 60, 115

Stefansson, Vilhjalmur 29
Stein, Auriel 28
Steller, Georg Wilhelm 88
Stuart, Revd John 88
Studies in Frontier History 34
Suess, Eduard 71
Sumatra 121
surveying 77, 79, 80–2, 109
  hydrographical 10–21, 122, 123
Switzerland 25, 87, 102–5
System of Value-creating Pedagogy 46, 47, 49–53

Tahiti 12–14, 19
Taiwan (Formosa) 51
Tasmania 13–14
taxonomy 91, 92, 96
teaching 43–53, 68–72, 104, 113, 114, 115
tectonics 93, 114
Terra Australis Incognita 11–21
textbooks 45, 50–1, 70, 114
Thailand 121, 122, 123
theatres 78
thematic maps 110–11
Tibet 37
Toda, Jogai 46–7
Tokyo Geographical Society 71
tonnage levy 2, 3, 4–5, 6
Tooke, Thomas 59
Torres Strait 14
Tours (Pennant) 87–98
Toynbee, Arnold 26, 34
trade routes 37, 120, 121, 123
Transit of Venus expedition 11–13, 18–19
travels 27–30, 69, 71, 72, 87–101; *see also* voyages
Turner, Frederick Jackson 35

Uchimura, Kanzo 45, 46
USA 28–31, 111–13, 115
Ussher, James 92–3
USSR 29, 62; *see also* Russia
utilitarianism 50, 52–3
Utley, Freda 31

value-creating pedagogy 46, 47, 49–53
Vidal de la Blache, Paul 114
Vienna University 69
Virginia Company 78
Vivó, Jorge A. 113
Voltaire 87
voyages 9–23, 120–3
vulcanism 93, 114

Waldron, Arthur 37
Wales 85–6, 87, 88–9, 95
Wallace, Henry 30

Wallace, Robert   59, 60
war   60, 61
Wheatley, Paul   37
Whitby   10, 21
White, Benjamin   87, 88
Willughby, Francis   86, 91
Wittfogel, Karl   31, 34
Wolman, Gordon   32
world geography   79–80, 85–98
World War II   29–31, 33, 53, 115
Wren, Christopher   79, 81

Yaganita, Kunio   45, 51

Zheng He   119–25
Zimmerman, E. A. W.   88, 91
zoology   63–4, 87–8, 89, 90–1
Zurich   87, 105

2. CUMULATIVE LIST OF BIOBIBLIOGRAPHIES

ADAIR, John (1660–1718)   *20*, 1–8
AL-BIRUNI (Abu'Rayhan Muhammad) (973–1054)   *13*, 1–9
AL-HASAN, see LEO AFRICANUS
AL-KINDI (801–873)   *17*, 1–8
ALMAGIA, Roberto (1884–1962)   *13*, 11–15
AL-MUQADDASI (c. 945–c. 988)   *4*, 1–6
ANCEL, Jacques (1882–1943)   *3*, 1–6
ANUCHIN, Dmitry Nikolaevich (1843–1923)   *2*, 1–8
APIANUS, Peter (1495 or 1501–1552)   *6*, 1–6
ARBOS, Philippe (1882–1956)   *3*, 7–12
ARDEN-CLOSE, Charles Frederick (1865–1952)   *9*, 1–13
ARMSTRONG, Terence Edward (1920–1996)   *18*, 1–9
ARQUÉ, Paul (1887–1970)   *7*, 5–9
ATWOOD, Wallace Walter (1872–1949)   *3*, 13–18
AUROUSSEAU, Marcel (1891–1983)   *12*, 1–8
BAKER, John Norman Leonard (1893–1971)   *16*, 1–11
BANSE, Ewald (1883–1953)   *8*, 1–5
BARANSKIY, Nikolay Nikolayevich (1881–1963)   *10*, 1–16
BATES, Henry Walter (1852–1892)   *11*, 1–5
BAULIG, Henri (1877–1962)   *4*, 7–17
BEAUFORT, Francis (1774–1857)   *19*, 1–15
BERG, Lev Semenovich (1876–1950)   *5*, 1–7
BERNARD, Augustin (1865–1947)   *3*, 19–27

BINGHAM, Millicent Todd (1880–1968)   *11*, 7–12
BLACHE, Jules (1893–1970)   *1*, 1–8
BLODGET, Lorin (1823–1901)   *5*, 9–12
BOBEK, Hans (1903–1990)   *16*, 12–22
BONNEY, Thomas George (1833–1923)   *17*, 9–16
BOSE, Nirmal Kumas (1901–1972)   *2*, 9–11
BOWEN, Emrys George (1900–1983)   *10*, 17–23
BOWMAN, Isaiah (1878–1950)   *1*, 9–18
BRATESCU, Constantin (1882–1945)   *4*, 19–24
BRAWER, Abraham Jacob (1884–1975)   *12*, 9–19
BRIGHAM, Albert Perry (1855–1929)   *2*, 13–19
BROOKS, Alfred Hulse (1871–1924)   *1*, 19–23
BROOKS, Charles Franklin (1891–1958)   *18*, 10–20
BROWN, Ralph Hall (1898–1948)   *9*, 15–20
BROWN, Robert Neal Rudmose (1879–1957)   *8*, 7–16
BRUCE, William Speirs (1867–1921)   *17*, 17–25
BUACHE, Philippe (1700–1773)   *9*, 21–7
BUJAK, Franciszek (1875–1953)   *16*, 23–30
BUSCHING, Anton Friedrich (1724–1793)   *6*, 7–15

CAMENA d'ALMEIDA, Pierre (1865–1943)   *7*, 1–4
CAPOT-REY, Robert (1897–1977)   *5*, 13–19
CAREY, Henry Charles (1793–1879)   *10*, 25–8
CAVAILLÈS, Henri (1870–1951)   *7*, 5–9
CHATTERJEE, Shiba P. (1903–1989)   *18*, 21–35
CHISHOLM, George Goudie (1850–1930)   *12*, 21–33
CHRISTALLER, Walter (1893–1969)   *7*, 11–16
CLARK, Andrew Hill (1911–1975)   *14*, 13–25
CLEMENTS, Frederic Edward (1874–1945)   *18*, 36–46
CODAZZI, Augustin (1793–1859)   *12*, 35–47
COLAMONICO, Carmelo (1882–1973)   *12*, 49–58
COLBY, Charles Carlyle (1884–1965)   *6*, 17–22
CONEA, Ion (1902–1974)   *12*, 59–72
COOK, James (1728–1779)   *20*, 9–23
COPERNICUS, Nicholas (1473–1543)   *6*, 23–9
CORNISH, Vaughan (1862–1948)   *9*, 29–35

CORTAMBERT, Eugène (1805–1881)   *2*, 21–5
COTTON, Charles Andrew (1885–1970)   *2*, 27–32
COWLES, Henry Chandler (1869–1939)   *10*, 29–33
CRESSEY, George Babcock (1896–1963)   *5*, 21–5
CUISINIER, Louis (1883–1952)   *16*, 96–100
CVIJIĆ, Jovan (1865–1927)   *4*, 25–32

D'ABBADIE, Antoine (1810–1897)   *3*, 29–33
DANA, James Dwight (1813–1895)   *15*, 11–20
DANTÍN-CERECEDA, Juan (1881–1943)   *10*, 35–40
DARWIN, Charles (1809–1882)   *9*, 37–45
DAVID, Mihai (1886–1954)   *6*, 31–3
DAVIDSON, George (1825–1911)   *2*, 33–7
DAVIS, William Morris (1850–1934)   *5*, 27–33
DE BRAHM, William Gerard (1718–1799)   *10*, 41–7
DE CHARPENTIER, Jean (1786–1855)   *7*, 17–22
DE MARTONNE, Emmanuel (1873–1955)   *12*, 73–81
DEE, John (1527–1608)   *10*, 49–55
DEMANGEON, Albert (1872–1940)   *11*, 13–21
DÍAZ COVARRUBIAS, Francisco (1833–1889)   *19*, 16–26
DICKEN, Samuel N. (1901–1989)   *13*, 17–22
DICKINSON, Robert Eric (1905–1981)   *8*, 17–25
DIMITRESCU-ALDEM, Alexandre (1880–1917)   *3*, 35–7
DION, Roger (1896–1981)   *18*, 47–52
DOKUCHAEV, Vasily Vasilyevich (1846–1903)   *4*, 33–42
DRAPEYRON, Ludovic (1839–1901)   *6*, 35–8
DRYER, Charles Redaway (1850–1927)   *11*, 23–6
DRYGALSKI, Erich von (1865–1949)   *7*, 23–9
DUNBAR, William (1749–1810)   *19*, 27–36

ERATOSTHENES (*c.* 275–*c.* 195 BC)   *2*, 39–43
EVEREST, Sir George (1790–1866)   *15*, 21–36
EYRE, Edward John (1815–1901)   *15*, 37–50

FABRICIUS, Johann Albert (1668–1736)   *5*, 35–9
FAIRGRIEVE, James (1870–1953)   *8*, 27–33
FAWCETT, Charles Bungay (1883–1952)   *6*, 39–46
FEDCHENKO, Alexei Pavlovich (1844–1873)   *8*, 35–8
FENNEMAN, Nevin Melancthon (1865–1945)   *10*, 57–68
FITZROY, Robert (1805–1865)   *11*, 27–33
FLEURE, Herbert John (1877–1969)   *11*, 35–51
FORBES, James David (1809–1868)   *7*, 31–7
FORMOZOV, Alexander Nikolayevich (1899–1973)   *7*, 39–46
FORREST, Alexander (1849–1901) and FORREST, John (1847–1918)   *8*, 39–43
FRANZ, Johann Michael (1700–1761)   *5*, 41–8
FRESHFIELD, Douglas William (1845–1934)   *13*, 23–31

GANNETT, Henry (1846–1914)   *8*, 45–9
GAVIRA MARTÍN, José (1903–1951)   *19*, 37–49
GEDDES, Arthur (1895–1968)   *2*, 45–51
GEDDES, Patrick (1854–1932)   *2*, 53–65
GEIKIE, Archibald (1835–1924)   *3*, 39–52
GERASIMOV, Innokentii Petrovich (1905–1985)   *12*, 83–93
GILBERT, Edmund William (1900–1973)   *3*, 63–71
GILBERT, Grove Karl (1843–1918)   *1*, 25–33
GILLMAN, Clement (1882–1946)   *1*, 35–41
GLACKEN, Clarence James (1909–1989)   *14*, 27–41
GLAREANUS, Henricus (1488–1563)   *5*, 49–54
GMELIN, Johann Georg (1709–1755)   *13*, 33–7
GOBLET, Yann-Morvran (1881–1955)   *13*, 39–44
GOODE, John Paul (1862–1932)   *8*, 51–5
GOYDER, George Woodroffe (1826–1898)   *7*, 47–50
GRADMANN, Robert (1865–1950)   *6*, 47–54
GRANÖ, Johannes Gabriel (1882–1956)   *3*, 73–84
GREELY, Adolphus Washington (1844–1935)   *17*, 26–42
GRIGORYEV, Andrei Alexandrovich (1883–1968)   *5*, 55–61
GUYOT, Arnold Henry (1807–1884)   *5*, 63–71

HASSERT, Ernst Emil Kurt (1868–1947)   *10*, 69–76
HAUSHOFER, Karl (1869–1946)   *12*, 95–106
HERBERTSON, Andrew John (1865–1915)   *3*, 85–92

HERDER, Johann Gottfried (1744–1803) *10*, 77–84
HETTNER, Alfred (1859–1941) *6*, 55–63
HIMLY, Louis-Auguste (1832–1906) *1*, 43–7
HO, Robert (1921–1972) *1*, 49–54
HÖHNEL, Ludwig von (1857–1942) *7*, 43–7
HOLMES, James Macdonald (1896–1966) *7*, 51–5
HOWITT, Alfred William (1830–1908) *15*, 51–60
HUGHES, William (1818–1876) *9*, 47–53
HUGUET DEL VILLAR, Emilio (1871–1951) *9*, 55–60
HULT, Ragnar (1857–1899) *9*, 61–9
HUTCHINGS, Geoffrey Edward (1900–1964) *2*, 67–71

IBN BATTUTA (1304–1378) *14*, 1–11
IGLÉSIES-FORT, Josep (1902–1986) *12*, 107–11
ILEŠIČ, Svetozar (1907–1985) *11*, 53–61
ISACHSEN, Fridtjov Eide (1906–1979) *10*, 85–92
ISIDA, Ryuziro (1904–1979) *15*, 61–74

JAMES, Preston Everett (1899–1986) *11*, 63–70
JOBBERNS, George (1895–1974) *5*, 73–6
JONES, Llewellyn Rodwell (1881–1947) *4*, 49–53

KANT, Edgar (1902–1978) *11*, 71–82
KANT, Immanuel (1724–1804) *4*, 55–67
KECKERMANN, Bartholamäus (1572–1609) *2*, 73–9
KELTIE, John Scott (1840–1927) *10*, 93–8
KENDREW, Wilfrid George (1884–1962) *17*, 43–51
KIM, Chong-ho (*c*. 1804–1866) *16*, 37–44
KINGSLEY, Mary Henrietta (1862–1900) *19*, 50–65
KIRCHOFF, Alfred (1838–1907) *4*, 69–76
KOMAROV, Vladimir Leontyevitch (1862–1914) *4*, 77–86
KRAUS, Theodor (1894–1973) *11*, 83–7
KROPOTKIN, Pyotr (Peter) Alexeivich (1842–1921) *7*, 57–62, 63–9
KRÜMMEL, Johann Gottfried Otto (1854–1912) *10*, 99–104
KUBARY, Jan Stanislaw (1846–1896) *4*, 87–9

LARCOM, Thomas Aiskew (1801–1879) *7*, 71–4
LATTIMORE, Owen (1900–1989) *20*, 24–42
LAUTENSACH, Hermann (1886–1971) *4*, 91–101

LEFÈVRE, Marguerite Alice (1894–1967) *10*, 105–10
LEICHHARDT, Friedrich (1813–1848?) *17*, 52–67
LEIGHLY, John (1895–1986) *12*, 113–19
LELEWEL, Joachim (1786–1861) *4*, 103–12
LENCEWICZ, Stanislaw (1899–1944) *5*, 77–81
LEO AFRICANUS (Al-Hasan ibn Muhammad al Wazzân az-Zayyâtî) (*c*. 1499–1550) *15*, 1–9
LEPEKHIN, Ivan Ivanovich (1740–1802) *12*, 121–3
LEVASSEUR, Emile (1828–1911) *2*, 81–7
LEWIS, William Vaughan (1907–1961) *4*, 113–20
LI DAOYUAN (*fl. c.* AD 500) *12*, 125–31
LINTON, David Leslie (1906–1971) *7*, 75–83
LLOBET I REVERTER, Salvador (1908–1991) *19*, 66–74
LOMONOSOV, Mikhail Vasilyevich (1711–1765) *6*, 65–70

MacCARTHY, Oscar (1815–1894) *8*, 57–60
McGEE, William John (1853–1912) *10*, 111–16
MACKINDER, Halford John (1861–1947) *9*, 71–86
MAGELLAN, Ferdinand (*c*. 1480–1521) *18*, 53–66
MAKAROV, Stepan Osipovich (1848–1904) *11*, 89–92
MAKIGUCHI, Tsunesaburo (1871–1944) *20*, 43–56
MALTHUS, Thomas Robert (1766–1834) *20*, 57–67
MARX, Karl (1818–1883) *19*, 75–85
MASON, Kenneth J. (1887–1976) *18*, 67–72
MATTHES, François Emile (1874–1948) *14*, 43–57
MAURY, Matthew Fontaine (1806–1873) *1*, 59–63
MAY, Jacques M. (1896–1975) *7*, 85–8
MEHEDINTI, Simion (1868–1962) *1*, 65–72
MELANCHTHON, Philipp (1497–1560) *3*, 93–7
MELIK, Anton (1890–1966) *9*, 87–94
MENTELLE, Edmunde (1730–1815) *11*, 93–104
MENTELLE, François-Simon (1731–1799) *11*, 93–104
MEURIOT, Paul (1861–1919) *16*, 45–52
MIHAILESCU, Vintila (1890–1978) *8*, 61–7

MILL, Hugh Robert (1861–1950)   *1*, 73–8
MILNE, Geoffrey (1898–1942)   *2*, 89–92
MITCHELL, Thomas Livingstone (1792–1855)   *5*, 83–7
MUELLER, Ferdinand Jakob Heinrich von (1825–1896)   *5*, 89–93
MUIR, John (1838–1914)   *14*, 59–67
MUNSTER, Sebastian (1488–1552)   *3*, 99–106
MUSHETOV, Ivan Vasylievitch (1850–1902)   *7*, 89–91
MYRES, John Linton (1869–1954)   *16*, 53–62

NAKANOME, Akira (1874–1959)   *20*, 68–76
NALKOWSKI, Waclaw (1851–1911)   *13*, 45–52
NANSEN, Fridtjof (1861–1930)   *16*, 63–79
NELSON, Helge (1882–1966)   *8*, 69–75
NEUSTRUEV, Sergei Semyonovich (1874–1928)   *8*, 77–80
NIELSEN, Niels (1893–1981)   *10*, 117–24

OBERHUMMER, Eugen (1859–1944)   *7*, 93–100
OBRUCHEV, Vladimir Afanas'yevich (1863–1956)   *11*, 105–10
O'DELL, Andrew Charles (1909–1966)   *11*, 111–22
OGAWA, Takuji (1870–1941)   *6*, 71–6
OGILBY, John (1600–1676)   *20*, 77–84
ORGHIDAN, Nicolai (1881–1967)   *6*, 77–9
ORMSBY, Hilda (1877–1973)   *5*, 95–7

PALLAS, Peter Simon (1741–1811)   *17*, 68–81
PARSONS, James Jerome (1915–1997)   *19*, 86–101
PARTSCH, Joseph Franz Maria (1851–1925)   *10*, 125–33
PAULITSCHKE, Philipp (1854–1899)   *9*, 95–100
PAVLOV, Alexsei Petrovich (1854–1929)   *6*, 81–5
PAWLOWSKI, Stanislaw (1882–1940)   *14*, 69–81
PENCK, Albrecht (1858–1945)   *7*, 101–8
PENNANT, Thomas (1726–1798)   *20*, 85–101
PERRON, Charles-Eugène (1837–1909)   *20*, 102–7
PETERMANN, August Heinrich (1822–1878)   *12*, 133–8
PHILIPPSON, Alfred (1864–1953)   *13*, 53–61
PITTIER, Henri-François (1857–1950)   *10*, 135–42

PLATT, Robert Swanton (1891–1964)   *3*, 107–16
PLEWE, Ernst (1907–1986)   *13*, 63–71
POL, Wincenty (1807–1872)   *2*, 93–7
POLO, Marco (1254–1324)   *15*, 75–89
POWELL, John Wesley (1834–1902)   *3*, 117–24
PRICE, Archibald Grenfell (1892–1977)   *6*, 87–92
PUMPELLY, Raphael (1837–1923)   *14*, 83–92

RAIMONDI DEL ACQUA, Antonio (1826–1890)   *16*, 80–7
RAISZ, Erwin Josephus (1893–1968)   *6*, 93–7
RATZEL, Friedrich (1844–1904)   *11*, 123–32
RAVENSTEIN, Ernst Georg (1834–1913)   *1*, 79–82
RECLUS, Elisée (1830–1905)   *3*, 125–32
RECLUS, Paul (1858–1941)   *16*, 88–95
REISCH, Gregor (*c.* 1470–1525)   *6*, 99–104
RENNELL, James (1742–1830)   *1*, 83–8
REVERT, Eugène (1895–1957)   *7*, 5–9
RHETICUS, Georg Joachim (1514–1573)   *4*, 121–6
RICHTER, Eduard (1847–1905)   *10*, 143–8
RICHTHOFEN, Ferdinand Freiherr von (1833–1905)   *7*, 109–15
RITTER, Carl (1779–1859)   *5*, 99–108
ROE, Frank Gilbert (1878–1973)   *18*, 73–81
ROMER, Eugeniusz (1871–1954)   *1*, 89–96
ROSBERG, Johan Evert (1864–1932)   *9*, 101–8
ROSIER, William (1856–1924)   *10*, 149–54
ROXBY, Percy Maude (1880–1947)   *5*, 109–16
RÜHL, Alfred (1882–1935)   *12*, 139–47
RUSSELL, Richard Joel (1895–1971)   *4*, 127–38
RYCHKOV, Peter Ivanovich (1712–1777)   *9*, 109–12

SALISBURY, Rollin D. (1858–1922)   *6*, 105–13
SÁNCHEZ GRANADOS, Pedro C. (1871–1956)   *20*, 108–18
SAUER, Carl Ortwin (1889–1975)   *2*, 99–108
SAWICKI, Ludomir Slepowran (1884–1928)   *9*, 113–19
SCHLÜTER, Otto (1872–1959)   *6*, 115–22
SCHMITHÜSEN, Josef (1909–1984)   *14*, 93–104
SCHMITTHENNER, Heinrich (1887–1957)   *5*, 117–21

Index

SCHRADER, Franz (1844–1924) *1*, 97–103
SCHWERIN, Hans Hugold von (1853–1912) *8*, 81–6
SCORESBY, William (1789–1857) *4*, 139–47
SEMËNOV-TYAN SHANSKIY, Pëtr Petrovich (1827–1914) *12*, 149–58
SEMËNOV-TYAN SHANSKIY, Veniamin Petrovich (1870–1942) *13*, 67–73
SEMPLE, Ellen Churchill (1863–1932) *8*, 87–94
SHALER, Nathaniel Southgate (1841–1906) *3*, 133–9
SHEN KUO (1033–1097) *11*, 133–7
SHIGA, Shigetaka (1863–1927) *8*, 95–105
SIBBALD, Robert (1641–1722) *17*, 82–91
SIEVERS, Wilhelm (1860–1921) *8*, 107–10
SION, Jules (1879–1940) *12*, 159–65
SMITH, George Adam (1856–1942) *1*, 105–6
SMITH, Wilfred (1903–1955) *9*, 121–8
SMOLENSKI, Jerzy (1881–1940) *6*, 123–7
SÖLCH, Johann (1883–1951) *7*, 117–24
SOLÉ I SABARÍS, Lluís (1908–1985) *12*, 167–74
SOMERVILLE, Mary (1780–1872) *2*, 100–11
SPENCER, Joseph Earle (1907–1984) *13*, 81–92
STAMP, Laurence Dudley (1898–1966) *12*, 175–87
STÖFFLER, Johannes (1452–1531) *5*, 123–8
STOKES, John Lort (1811–1885) *18*, 82–93
STRZELECKI, Pawel Edmund (1797–1873) *2*, 113–18

TAMAYO, Jorge Leonides (1912–1978) *7*, 125–8
TANSLEY, Arthur George (1871–1955) *13*, 93–100
TATISHCHEV, Vasili Nikitich (1686–1750) *6*, 129–32
TAYLOR, Thomas Griffith (1880–1963) *3*, 141–53
TEILHARD DE CHARDIN, Pierre (1881–1955) *7*, 129–33
TELEKI, Paul (1879–1941) *11*, 139–43
TERAN-ALVAREZ, Manuel de (1904–1984) *11*, 145–53
THOMPSON, David (1770–1857) *18*, 94–112
THORNTHWAITE, Charles Warren (1899–1963) *18*, 113–29
TILLO, Alexey Andreyevich (1839–1900) *3*, 155–9
TOPELIUS, Zachris (1818–1898) *3*, 161–3
TORRES CAMPOS, Rafael (1853–1904) *13*, 102–7

TOSCHI, Umberto (1897–1966) *11*, 155–64
TROLL, Carl (1899–1975) *3*, 111–24
TULIPPE, Omer (1896–1968) *11*, 165–72

ULLMAN, Edward Louis (1912–1976) *9*, 129–35

VALLAUX, Camille (1870–1945) *2*, 119–26
VALSAN, Georg (1885–1935) *2*, 127–33
VAN CLEEF, Eugene (1887–1973) *9*, 137–43
VAVILOV, Nikolay Ivanovich (1887–1943) *13*, 109–16, 117–32
VERNADSKY, Vladimir Ivanovich (1863–1945) *7*, 135–44
VICENS VIVES, Jaume (1910–1960) *17*, 92–105
VIDAL DE LA BLACHE, Paul (1845–1917) *12*, 189–201
VILA I DINARES, Pau (1881–1980) *13*, 133–40
VIVEN DE SAINT-MARTIN, Louis (1802–1896) *6*, 133–8
VOLZ, Wilhelm (1870–1959) *9*, 145–50
VOYEIKOV, Alexander Ivanovich (1842–1916) *2*, 135–41
VUIA, Ramulus (1881–1980) *13*, 141–50
VUJEVIC, Pavle (1881–1966) *5*, 129–31

WAIBEL, Leo Heinrich (1888–1951) *6*, 139–47
WALLACE, Alfred Russel (1823–1913) *8*, 125–33
WANG YUNG (1899–1956) *9*, 151–4
WARNTZ, William (1922–1988) *19*, 102–7
WARD, Robert DeCourcy (1867–1931) *7*, 145–50
WATSON, James Wreford (1915–1990) *17*, 106–15
WELLINGTON, John Harold (1892–1981) *8*, 135–40
WEULERSSE, Jacques (1905–1946) *1*, 107–12
WILKES, Charles (1798–1877) *15*, 91–104
WISSLER, Clark (1870–1947) *7*, 151–4
WOOLDRIDGE, Sidney William (1900–1963) *8*, 141–9
WU SHANG SHI (1904–1947) *13*, 151–4

XU HONGZU (1587–1641) *16*, 31–6

YAMASAKI, Naomasa (1870–1928) *1*, 113–17

ZHENG HE (1371–1433) *20*, 119–25